SCIENCE AND SOCIAL CHANGE IN BRITAIN AND EUROPE 1700–1900

SCIENCE AND SOCIAL CHANGE IN BRITAIN AND EUROPE

1700–1900

Colin A. Russell

St. Martin's Press New York

St. Martin's Press, Inc., 175 Fifth Avenue, New York, NY 10010
Printed in Hong Kong
Published in the United Kingdom by The Macmillan Press Ltd.
First published in the United States of America in 1983

ISBN 0-312-70239-6

Library of Congress Cataloging in Publication Data
Russell, Colin Archibald.
Science and social change in Britain and Europe, 1700–1900.

Bibliography: p.
Includes indexes.
1. Science–Social aspects–Great Britain–History.
2. Science–Social aspects–Europe–History. I. Title.
Q175.52.G7R87 1983 303.4′83 83-11021
ISBN 0-312-70239-6

Contents

List of Figures

Figures

List of Plates

The author and publisher wish to thank the following illustration sources: The Science Museum – 1; The British Library – 3; The Evangelical Library – 4; The British Museum – 5; The Mansell Collection – 8, 9; The Royal Institution – 10; The Archives of Imperial College of Science and Technology, London – 17; The Victoria and Albert Museum – 21.

TO MY FATHER
AND IN MEMORY
OF MY MOTHER

General Editor's Preface

SINCE the Second World War there has been a massive expansion in the study of economic and social history generating, and fuelled by, new journals, new academic series and societies. The expansion of research has given rise to new debates and ferocious controversies. This series proposes to take up some of the current issues in historical debate and explore them in a comparative framework.

Historians, of course, are principally concerned with unique events, and they can be inclined to wrap themselves in the isolating greatcoats of their 'country' and their 'period'. It is at least arguable, however, that a comparison of events, or a comparison of the way in which different societies coped with a similar problem – war, industrialisation, population growth and so forth – can reveal new perspectives and new questions. The authors of the volumes in this series have each taken an issue to explore in such a comparative framework. The books are not designed to be path-breaking monographs, though most will contain a degree of new research. The intention is, by exploring problems across national boundaries, to encourage students in tertiary education, in sixth-forms, and hopefully also the more general reader, to think critically about aspects of past developments. No author can maintain strict objectivity; nor can he or she provide definitive answers to all the questions which they explore. If the authors generate discussion and increase perception, then their task is well done.

Clive Emsley

Preface

STUDIES of science in its social context have become so fashionable in recent years that historical analyses of internal developments within the sciences have become comparatively rare. Combined with this trend has been a proliferation of highly specialised studies of great interest to scholars in the field, but unfortunately inaccessible to a wider public. In this respect the History of Science is in some danger of becoming a victim of its own new-found professionalism. For some years one has searched publishers' lists anxiously, but in vain, for works on the present subject that could reasonably be expected to serve as a general introduction or as student texts. It was this gap in the literature, more than any other consideration, that goaded a rather reluctant author into producing the present book. It is intended for any who are seriously concerned to study the historical dimensions of the relations between science and society, whether approaching the subject from general history or from science itself. Some ten years' experience at the Open University has persuaded me that such an enterprise, though apparently foolhardy, is indeed possible, and that the common ground between historians and scientists is surprisingly great.

The theme of this book is the multitude of ways in which science has reflected, influenced and indeed become part of the wider phenomena of social change. Considerations of length have limited the period to a mere two centuries, and also concentrated attention largely on one country, Britain, though attempts have been made to place the British experience in a wider international context. An area that has been covered far less thoroughly than I should have wished is the impact of science on living and working conditions, and on the means of destroying or prolonging life. Adequate treatment of these themes would require another book-length study.

Warm thanks are offered to the staffs of many institutions including the Public Record Office, the British Library, The Wellcome Historical Medical Library, Bedfordshire County Library, Bedford Central Library, Lancaster Public Library, Newcastle City Library, and the Libraries of the Open University, the University of London, the Royal Society of Chemistry, Guildhall, London and Preston Polytechnic. I am grateful to the Royal Institution for permission to cite the manuscript notebook of Thomas Webster and to Mr George Rolleston for permission to quote from a Davy letter in his possession; also to the Harvester Press for permission to make extensive citations from *The Newtonians and the English Revolution 1689 to 1720*, by Margaret Jacob. Acknowledgement for illustrations will be found on page x.

It will be apparent that my debt to other scholars is enormous, especially to those represented in the footnotes. I can only hope that none have been misrepresented, especially when I have disagreed with their conclusions. One work above all others has been a catalyst, inspiration and (in some respects) a model. Its excellence has at the same time relieved me of the necessity of covering much of the same ground: Donald Cardwell's *The Organisation of Science in England,* 2nd edn (London: Heinemann, 1972).

I am most grateful to several colleagues who have read and criticised parts of the book: Dr John H. Brooke, Dr Robert Fox and Dr Gerrylynn K. Roberts, and especially to the Editor of this series, Mr Clive Emsley, whose comments on the whole manuscript have been invaluable. Needless to say the responsibility for what follows is mine alone.

My thanks must also go to Dr Eamann Duffy for allowing me to see a manuscript version of his forthcoming paper on 'The Whiston circle and heterodoxy'; to Dr Robert Fox for showing me an as yet unpublished essay on the profession of science in nineteenth-century France; and to Dr Jim Moore for a pre-publication copy of his paper on the burial of Charles Darwin in Westminster Abbey.

Nor must I forget the often passionate debates with colleagues at the Open University where many of these issues first became sharpened in my own mind. To them all I am very grateful, as I am to the Open University for a year of study leave during which this book took shape. My gratitude also goes to our two secretaries in the History of Science group, Mrs Patricia Dixon and Mrs Stephanie Lansdown, who between them have coped valiantly with innumerable drafts of chapters and the collection and collation of appropriate illustrations.

Finally, I have to thank my family who have had to live for far too long with my preoccupation with the book but who have nevertheless exhibited monumental patience and understanding.

COLIN A. RUSSELL 1982
The Open University
Milton Keynes

Abbreviations

Amer. Hist. Rev.	*American Historical Review*
Ann. Sci.	*Annals of Science*
B. J. Hist. Sci.	*British Journal for the History of Science*
Berichte	*Berichte der Deutschen chemischen Gesellschaft*
Biol. J. Linnaen Soc.	*Biological Journal of the Linnaen Society*
Bull. Inst. Hist. Res.	*Bulletin of the Institute of Historical Research*
Chem. Age	*Chemical Age*
Chem. and Ind.	*Chemistry and Industry*
Chem. Brit.	*Chemistry in Britain*
Chem. News	*Chemical News*
DNB	*Dictionary of National Biography*
DSB	*Dictionary of Scientific Biography*
Econ. Hist. Rev.	*Economic History Review*
Hist. J.	*Historical Journal*
Hist. Sci.	*History of Science*
Hist. Stud. Phys. Sci.	*Historical Studies in the Physical Sciences*
J. Chem. Educ.	*Journal of Chemical Education*
J. Hist. Ideas	*Journal of the History of Ideas*
J. Roy. Inst. Chem.	*Journal of the Royal Institute of Chemistry*
J. Roy. Soc. Arts	*Journal of the Royal Society of Arts*
J. Soc. Bibliography Nat. Hist.	*Journal of the Society for the Bibliography of Natural History*
J. Soc. Chem. Ind.	*Journal of the Society of Chemical Industry*
M. & B. Lab. Bull.	*May and Baker Laboratory Bulletin*

Med. Hist.	*Medical History*
Notes and Records	*Notes and Records of the Royal Society*
Oxford Rev. Educ.	*Oxford Review of Education*
Proc. Chem. Soc.	*Proceedings of the Chemical Society*
Proc. Inst. Chem.	*Proceedings of the Institute of Chemistry*
Proc. Roy. Inst.	*Proceedings of the Royal Institution*
Proc. Roy. Soc. Edinburgh	*Proceedings of the Royal Society of Edinburgh*
Qu. J. Chem. Soc.	*Quarterly Journal of the Chemical Society*
Qu. J. Geol. Soc.	*Quarterly Journal of the Geological Society*
Rep. Brit. Assoc. Adv. Sci.	*Reports of the British Association for the Advancement of Science*
Soc. Stud. Sci.	*Social Studies of Science*
Trans. Newcastle Chem. Soc.	*Transactions of the Newcastle Chemical Society*
Trans. Newcomen Soc.	*Transactions of the Newcomen Society*
Trans. Roy. Geol. Soc. Cornwall	*Transactions of the Royal Geological Society of Cornwall*
Trans. Roy. Soc. Edinburgh	*Transactions of the Royal Society of Edinburgh*
Trans. Tyne Chem. Soc.	*Transactions of the Tyne Chemical Society*

SUPPLEMENTARY LIST OF ABBREVIATIONS

Adv. Sci.	*Advances in Science*
B. J. Educ. Studies.	*British Journal of Educational Studies*
Comptes Rendus	*Comptes Rendus Lebdonadaires des Séances de l'Académie des Sciences (Paris)*
Eng. Hist. Rev.	*English Historical Review*
J. Brit. Astronom. Assoc.	*Journal of the British Astronomical Association*
J. Soc. Arts.	*Journal of the Society of Arts*
J. Soc. Hist.	*Journal of Social History*
J. Textile Inst.	*Journal of the Textile Institutions*
Proc. Roy. Soc.	*Proceedings of the Royal Society of London*
Trans. Beds. Nat. Hist. Soc.	*Transactions of the Bedfordshire Natural History Society*

1. Approaching the Problem

IN the two centuries from 1700 to 1900 the face of science was transformed almost beyond recognition. From the point of view of a profoundly changed conception of nature it was not, perhaps, so revolutionary a change as had taken place within the seventeeth century itself, but in terms of visible scientific achievement, a transformation of popular attitudes towards science and the impact of science on society, it was entirely without precedent.

In 1700 Isaac Newton was looking back, somewhat wearily, on a lifetime of stupendous accomplishment culminating in one of the greatest scientific works of all time, the *Principia* of 1686. Now Master of the Mint he had 27 years ahead of him as elder statesman of science and (from 1703) President of the Royal Society. In 1900 a young student was completing his education at Zurich and about to produce his first paper: Albert Einstein. The two centuries in between were to witness the rise and fall of one of the great generalised systems of the human intellect, Newtonian physics. Victorian thermodynamics as well as Einstein's relativity theories were responsible for its overthrow, or perhaps one should say its incorporation into a far wider scheme. This was not, of course, the only revolution in physical science. By 1900 that ancient fancy once sanctified by Newton, the all-pervasive ether, was ready to be laid to rest. It was the last survivor of a host of imponderable fluids that were gradually eliminated during this long period. Techniques in astronomy unknown in 1700 were regularly in use 200 years later, including photography and spectroscopy, while telescope size and design were greatly advanced. As a result the universe of 1900 was immeasurably vaster and more complex than that of 1700. Early in the eighteenth century Flamsteed's Star Catalogue mapped 2884 stars; the Henry Draper Catalogue published early in this century gave data for about 225,000 of the most luminous.

Notions of electricity as a curious by-product from rubbing amber were in 200 years replaced by recognition of current and galvanic electricity and an electromagnetism that could be understood in terms of a unified field theory which embraced also light and other forms of radiation. The eternally existing, 'hard, massy' atoms that had sufficed from the ancients to Newton had, by 1900, been replaced by particles that could disintegrate in radioactivity and were very soon to be shown to consist largely of empty space. Chemical elements had been defined by Boyle in the late seventeenth century, though he was able to identify none. A table printed in 1900 listed sixty-nine of the ninety-odd naturally occurring elements. The legacy from the alchemists of transmutation was still to be found in 1700, though it disappeared soon afterwards, only to find an apparent reincarnation just before 1900 with the discovery of radioactivity. During this period, that notable though erroneous unifying principle of chemistry, the theory of phlogiston, had come and gone. In 1700 only fourteen or fifteen pure organic compounds had been recognised and studied; by 1900 their number was about 100,000.

The idea of geology as a science in 1700 is a crude anachronism. At the end of the period it was sufficiently prestigious for the great Lord Kelvin to hail it as a 'branch of physical science' and a 'grand subject'. In biological studies it is remarkable how little had changed in the two millennia before 1700, Harvey and Ray notwithstanding. But by 1900 the static, hierarchical world of living nature was replaced by one of ceaseless change, the supposedly immutable species being merely transient stages in the inexorable if incomprehensible processes of evolution. And well before that date not a few were speculating that biological phenomena were 'nothing but' the operations of laws of chemistry and physics upon 'dead matter'.[1]

Nowhere could the scale of these cataclysmic changes be more clearly seen than in Britain. History is far more than the study of 'great men', but merely to mention names is to suggest something of the remarkable progress of British science: Newton, Bradley, the Herschels, Lockyer; Priestley, Black, Dalton, Davy, Faraday, Frankland, Crookes; Cavendish, Tyndall, Joule, Kelvin, Rutherford; Ray, Hales, Hutton, Lyell, Darwin, Wallace, Huxley, and so on. It is precisely at this point that one catches a first glimpse of science in its social context. The very fact that it is meaningful to talk of 'British science', 'German chemistry', 'French physics' etc., is clear evidence of cultural influences upon science. Since it is much harder to do this with modern science (at least in the West) one may fairly infer that the internationalisation of science owes much to the massive social changes that have overtaken the world in recent years.

The eighteenth and nineteenth centuries witnessed a series of convulsive political revolutions in Europe and America that in their earlier phase have been called 'the Western Atlantic Revolution'.[2] In Europe these and other wars led to more than a redrawing of the political map, the breaking of old alliances and the forging of new ones. The resultant redistribution of wealth and power and the emergence of new élites in most countries were accompanied by civil strife that often involved the new industrial working classes. In Britain, as elsewhere, the process of industrialisation, accompanied by urbanisation, constituted a major social discontinuity with the past. In this country the changes were particularly significant and earlier than most because here the Industrial Revolution had its historic roots. Two other extremely important trends, though neither originated within this period, were the growth of capitalism and the increase in secularism.

Given the spectacular growth of science, it is tempting to relate this to the wider panorama of social change occurring at the same time. One might, for example, wish to enquire whether social changes had led to developments in science, or vice versa:

$$\text{SOCIETY} \rightleftarrows \text{SCIENCE}$$

Traditionally historians of science have been rather more

interested in the questions represented by the upper arrow than those associated with the lower, partly because in the former case science is the dominant element. Before these and other models can be employed, however, one needs to deal with an objection that is sometimes levelled against this formulation of the problem. It has been argued that it assumes that science and society are two autonomous entities, each capable of existing independently of the other. Even to write down the problem in these terms, therefore, begs the central question. An extreme version of this thesis is the aphorism 'science *is* social relations'.[3]

To assertions of that kind one may simply reply that there is more than one way of looking at science, and to say that it is 'nothing but' a social phenomenon is just as much an assumption as the older view that science operates quite independently of social constraints. If for the moment we assume that scientific *activity* can be defined in purely societal terms (leaving aside the question of scientific *knowledge*), we can redefine the problem as how a set of social activities which we might call 'science' can be related to other sets which we label otherwise.

Sooner or later, however, we have to come to terms with what we mean by science in its broadest sense. This is particularly necessary in historical enquiries because it would be rash in the extreme to suppose that the meaning of 'science' did not vary with time. In his book *The Nature of Science*, David Knight begins with a view of science as 'a process of thinking about nature, of talking about nature, and of interrogating and using nature. That is, we shall describe science as an intellectual, a social, and a practical activity. It is only if we follow such a broad and comprehensive road that we can do justice to the complexity of science.'[4] This excludes the common view of science as a collection of facts, and quite legitimately avoids consequent questions as to the objectivity or otherwise of those facts. With this understanding it becomes perfectly acceptable to enquire how science and society may have interacted in the past. Several models have been proposed, and each may be helpful to the historian in suggesting lines of enquiry.

First we have what may be called the zero-interaction

model, at least in relation to the possibility that science may be affected by society. In 1977 the *New Scientist* carried an article by Professor M. Hammerton which designated as 'a fashionable fallacy' the notion that scientific advances are ever determined by 'social and economic factors'.[5] Pointing out, quite rightly, that a literal interpretation of this view 'is not only false but absurd', he cited the imaginary possibility of a society in which the second law of thermodynamics[6] was denied and in which it did not even apply. Yet if science is entirely socially conditioned this is theoretically possible. Since no one is prepared to make that deduction he limits the proposition to its weaker, non-literal meaning, according to which society *may* condition scientific thinking. Presumably this would only be within the constraints imposed by nature itself, so the previous *reductio ad absurdum* would not apply. But for this to be believed, proof must be produced by the historians, and he concluded that no 'strong evidence' exists. A somewhat similar position was taken by D. W. F. Hardie in objecting to the argument that, because certain men shared an interest in science with (say) certain political or religious views, some direct causal connection may be assumed between the one and the other. He observed, 'In democratic societies at least to assert guilt by association is generally considered bad law; finding significance by association, and association alone, is equally bad historical procedure'.[7]

It was on this crucial historical point of evidence that the Hammerton view was effectively undermined. Apart from the more hysterical rejoinders was a temperate response by John Durant who indicated that just because the 'strong evidence' *did* exist, historians were unable to close their eyes to 'the need to interpret science in its cultural context'. To conclude 'that science develops in some sort of cultural vacuum, totally isolated from other human interests, is a classic case of jumping out of the frying pan into the fire'.[8] It is in fact no more reprehensible to argue that social and economic influences 'must have been' important than, with Hammerton, to assert they can't have been. Fashion apart, mounting evidence does compellingly point to a rejection of this model.

A second model, espoused once by hard-line Marxists, is to

propose that a necessary connection exists between the economic needs of industry and the formation of scientific theories. It is classically associated with the efforts of a Soviet delegation to the Second International Congress of the History of Science, held in London in 1931. Most famous (or notorious, according to one's point of view) was a paper by Boris Hessen, a physicist generally supposed to have disappeared during a Stalinist purge shortly afterwards. His essay, 'The social and economic roots of Newton's *Principia*' purported to see Newton's scientific work developing out of his interests in mining, artillery, navigation and other economically important activities.[9] Based largely on evidence from one youthful letter, the analysis is now generally regarded as simplistic, crude, even naïve. Many other efforts were made to give credibility to this economic model, but with one or two minor exceptions, it simply does not have the necessary facts to support it. As Morris Berman has said, 'Thirties Marxism . . . is not a difficult target to attack',[10] and few Marxist writers today would claim otherwise. A critique of Hessen's essay by G. N. Clark observed that Hessen failed to realise that at least five other areas of human activity could influence science as well as economic factors: war, medicine, arts, religion and 'the disinterested search for truth'.[11] Yet when even this is conceded for the late seventeenth century, the coming of the Industrial Revolution 100 years later still poses acute problems as to the relation between industrial needs and science, and Arnold Thackray has pointed out how several non-Marxist historians (such as Ashton and Halevy) have concluded that scientific theory in certain cases stemmed from industrial practice.[12] What they seem to be referring to is the process by which scientific theories are constructed rather than their cognitive content. At the very least such possibilities should be entertained, whether or not one may be profoundly unimpressed by Marxist ideology and historiography in general.

A third model is a more subtle one, and relies on the neo-Marxist concept of 'hegemony'. Berman[13] has sought to explain certain characteristics of early nineteenth-century British science in terms of the class-struggle. The ruling class seeks to establish and maintain a 'hegemony', or cultural supremacy, and thus to stamp its imprint on the whole of society, science included. Hence the practice of science will reflect values that originate in areas remote from it. The amateurism of British science at the time is seen as a reflection of aristocratic ideals. So also an analysis of membership and leadership of scientific institutions may reveal hidden cultural values of a class that wishes to govern or to go on governing. Much recent work has in fact demonstrated some extremely interesting cultural 'networks' and suggests science to have been a means to a political end. It appears that in many cases the practice of science may have been affected by local power-struggles, and in some cases it certainly was. The question of scientific institutions thus becomes invested with a new significance (which is why many of them figure prominently in this book). If such institutions do reflect cultural values – and it is hard to see how they could not – we next need to know how those values came to affect the science being pursued. We should not limit our enquiry to specialist scientific institutions as such, but must consider a wide range of issues from education to government funding.

If it be conceded that the practice of science owes much to its social context, it does not necessarily follow that scientific laws and theories share in the debt. Perception of generalisations in science is a complex matter. It may depend on scientific practice, so that the rate at which laws etc. are perceived may have some dependence on social context. But in the last analysis what is perceived will depend on what is 'there' in nature to be seen. An inverse cube law, instead of an inverse square law, for gravitational attraction, is as inconceivable in any social context as Hammerton's imaginary society in which thermodynamic laws did not hold and kettles were boiled by standing them on icebergs. Fashionable trends in the sociology of knowledge frequently ignore this most elementary of facts (and justifiably incur the wrath of many of those who actually practise the science). It is perfectly true that formation and propagation of scientific beliefs may be demonstrably a function of the milieu in which they appear, but it is wholly illogical to say therefore they are 'nothing but' a social construct. Some writers who have attempted to expose 'the myth of value-free science' have most enjoyed themselves when assailing scientific *knowledge*, for here is one of the last bastions of alleged objectivity.[14] It is a great pity that their fulminations (which are no more value-free than the science they attack) should have diverted attention from what is in effect a fourth model for the relation between science and society, this is that scientific theories *may* (not will) express cultural values in their origins and form as well as in their transmission and use. Theories of the ether, for instançe, were at times very culture-laden.[15]

A historian must be very alive to this kind of possibility. Theories and even laws do come about in strange ways. But their fate will depend on their conformity with experiments and observations of natural phenomena. The German chemist Kekulé, credibly said to have arrived at two crucial ideas in organic chemistry through reveries or dreams, observed: 'Let us learn to dream, gentlemen, but let our dreams be put to the test of a waking understanding',[16] and that involves interrogation of nature. Yet even here there is room for the possibility of cultural influences, generally unperceived at the time. A theory is verified, or falsified, by reference to the facts. But these facts have to be preselected – one cannot use all the empirical data available. Should a study of bird migration, for example, take into account terrestrial magnetism? For many years it did not occur to biologists to ask. That was the province of the physicist, and the division between biologists and physicists is nothing if not a cultural and social one. Yet eventually the earth's magnetism was seen to be crucial.

It is thus possible to envisage at least three possible ways for social forces to affect science: the crude economic model of early Marxism, the more subtle one of hegemony (which relates primarily to the practice of science), and a further model which partakes of certain characteristics of the other two and which posits a cognitive content for science that is susceptible to cultural influences of all kinds while still being conformable to nature. To these three, and the zero-interaction model, may be added two others which both relate to an 'influence' in the opposite direction and address the question: 'how does science affect society?'

Most obviously science can affect society by its practical

application. Whether in improving industrial processes, military weapons or medical care, today the effects of science are clear enough. But we should be warned. With hindsight it is possible to assign far too much weight to the social influence of science in this sense. The problem is exacerbated by claims on behalf of science that sought for some quite specific reason to exalt its status and therefore to exaggerate its achievements. If a distinction is made between science and technology it becomes apparent that many of the social upheavals in the eighteenth and nineteenth centuries originated with the latter, not the former. Thus the growth of the railways, the development of steam power in its early days and the mechanisation of the textile industry owed little to the systematic exploration of nature that is the hallmark of science, but much to ingenuity, assiduous tackling of a problem in a pragmatic and empirical way, and to the 'useful arts' rather than the abstract sciences. The distinction needs further elaboration, of course, and it is not an absolute one. Its crucial feature is that it leaves open the possibility that technology might be a direct product of scientific activity. Only towards the end of our period can science be positively identified as the mainspring for those developments in technology that demonstrably and immediately led to social change (see Chapter 14). One example must suffice. The spread of gas-lighting in the 1820s and afterwards had several direct results: schooling was possible later on winter days than hitherto, it has been argued that crime was drastically reduced in streets lit by gas, and industry was able to work round the clock. Yet gas lighting emerged from a combination of accidental discoveries, trial-and-error experiments and good showmanship. It owed little to science and was in fact opposed as dangerous (or impossible) by many of the scientific establishment.[17] On the other hand the incandescent electric lamp arose in a much more recognisably scientific milieu, after much carefully planned experimentation, and utilised ideas and techniques from several other areas of science. Yet the immediate social consequences of electric lighting were arguably less important than those from lighting by gas.

The possibility that industrial practice might be a direct consequence of scientific input is the opposite of the usual

Marxist thinking (in which the direction of 'influence' is reversed). As Thackray points out, the 'Marxian position is logically distinct from, though reconcilable with, the belief that the innovations in the Industrial Revolution were dependent on scientific expertise. Either, both, or neither position may be correct. Marxian orthodoxy has favoured the first and ignored the second.'[18] That, however, is no reason for continuing to ignore it any longer. But there is another way in which science may help to mould society and, in the case of Manchester, Thackray clearly regards this as the more important.

Science, according to this sixth model, may become 'a mode of cultural self-expression by a new social class', and may be adopted for many reasons which will naturally vary with circumstances. It will mean that, if successfully used, science will minister to the social needs of those whom Thackray calls 'marginal men', confirming them as new leaders of society and increasing their confidence; it could produce a measure of 'social control' by a variety of mechanisms; it might even involve a slow erosion of established authority and its replacement by more radical elements. Science in this way could be an ideological weapon for both the political left and right. It could be used to bolster established religion or to undermine it. But for any of these things to happen science must already have been widely received as beneficial, or otherwise there would be no point in using it. That means it must be commonly *thought* to have conferred practical benefits on mankind.

The relations between this model and the one based on hegemony must be clear, and in fact they differ only in emphasis, the latter stressing more the way in which the science is affected. If there is any validity in any of these proposals it must now be obvious that all kinds of mutual influences are possible, and a complex feedback situation can arise. Instead of our simple relation:

$$\text{SOCIETY} \rightleftarrows \text{SCIENCE}$$

it now becomes possible to envisage something rather more elaborate:

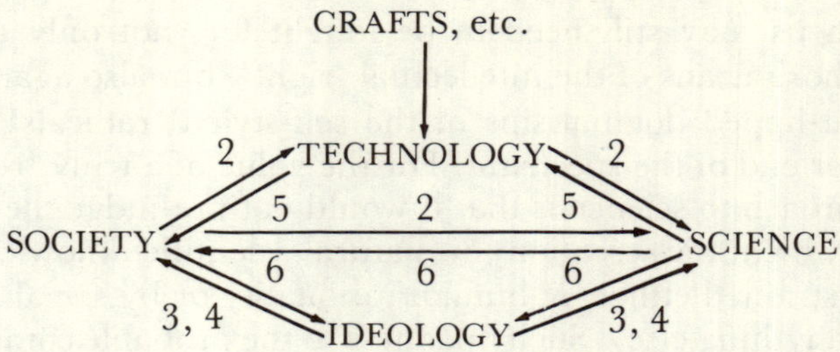

(numbers refer to models discussed earlier in the text)

All this, to a historian, is very theoretical, and quite valueless until individual situations are subjected to detailed enquiries. And so it should be. But to be alerted to possible relationships may enable them to be identified rather more easily, and to realise the probable complexity of a situation is to serve as a constant warning against commitment to any one ideology that purports to offer *the* key to understanding the social history of science. When Arthur Schuster and Arthur Shipley wrote their *Britain's Heritage of Science* they chose to preface it with the words 'This book does not pretend to establish any thesis'.[19] Neither, in any narrow sense, does the present one. But it does contend for a profoundly important relationship between science and society, of which the preceding discussion offers the merest hint.

It remains, perhaps, to add a note of caution. An important emphasis in the history of science has been provided by members of the Science Studies Unit of Edinburgh University – the 'Edinburgh School'. In a recent collection of essays,[20] *Natural Order: Historical Studies of Scientific Culture*, two members of the Group have urged a 'more relaxed and naturalistic' approach to the history of science in which the rhetoric of science as liberation from non-rational superstition is replaced by an openness to its faults, its dangers and its social conditioning. In an essay review of this work, Martin Rudwick several times reiterates the need for convincing historical evidence to back up assertions connecting science and society, and adds these remarks which serve as an admirable start to the present historical quest:[21]

The 'relaxed' approach to science that they claim is already with us may still need to be fought for, not only against authoritarians of the intellectual 'right', but also against the tight-lipped dogmatisms of the self-styled 'radicals' at the other end of the spectrum. For the value of a truly 'relaxed' approach to science is that it would not pre-judge the extent to which the creation of natural-scientific knowledge is constrained either by human rationality or by social life (or by anything else). Such openness to the probable complexity of scientific activity, like other forms of liberalism, may be unfashionable in the present harsh political climate; but then fashion has never been a desirable guide in any sphere of life.

2. The Shape of Enlightenment Science

2.1 PROLOGUE

BETWEEN the Glorious Revolution in England in 1688/9 and the French Revolution lay the century of the so-called 'Enlightenment'. To refer to it like that is, of course, to draw attention exclusively to its cultural and intellectual character, and to ignore the political, military and technical achievements of the time. It may also be to assent to the oddly optimistic assessment of their own age by many who saw little contradiction between its grandiose pretensions and what we should see as the sterility of much of its thought. However this is a convenient label, if only because it tells us how the eighteenth century often regarded itself. And for that attitude science was in no small measure responsible.

The Enlightenment also falls between two other so-called 'Revolutions'. The Scientific Revolution is a phrase sometimes used to encapsulate the train of events from the *De Revolutionibus Orbium Coelestium* of Copernicus (1543), to Newton's *Philosophiae Naturalis Principia Mathematica* (1687), via Galileo and Kepler, and including such events as William Gilbert's pioneering researches into magnetism (1600), Harvey's discovery of the circulation of the blood (1628), the work of Robert Boyle in chemistry and of Robert Hooke in microscopy, the mathematical advances of Leibniz and

Descartes, and a vast amount of scientific observation, with experiments conducted under conditions so controlled that repeatability became a realisable goal. Within the seventeenth century this greatly expanded sphere of human activity became institutionalised, especially in England and France with the creation of the Royal Society of London (1662) and the Académie Royale des Sciences (1666) on the basis of earlier Italian models. Scientific communication entered a new dimension with journals like the *Philosophical Transactions of the Royal Society* (1665), embodying the Society's preference for 'the language of artizans countrymen and merchants, before that of wits or scholars'.[1] Elsewhere in Europe scientific periodical literature appeared for the first time, including the short-lived French *Journal des Scavants* (which just predated *Philosophical Transactions*), and the *Acta Eruditorum* of Leipzig (1682).

If the Scientific Revolution ushered in a new kind of scientific organisation, at a more profound level it initiated a view of the universe that, for all its affinities with the past in points of detail, marked a vast and unprecedented disjunction in human thought (and, in the long term, human action). Partly it was a question of scale and relativity. Through the microscopes of Hooke whole new worlds of intricate beauty were revealed to the astonished observers. Yet never at the greatest magnification was there any trace of the legendary atoms of the Greeks – the ultimate particles of things. The fine structure of matter was infinitely more complex than had been imagined. At the other end of the optical scale the telescopes of Galileo and his successors had revealed unsuspected wonders in the heavens: sunspots, the satellites of Jupiter, the phases of Venus, multitudes of new stars, and so on. But the very notion of the Copernican Universe, supposedly corroborated by these discoveries, implied so much more. Gone was the cosy world of popular cosmology with man's home at the centre of his universe; more seriously, since man's displacement from the centre of a vast universe was comparatively trivial in extent (his place merely being taken by his relatively near neighbour the sun), he now lived on simply another planet whose unique status could no longer be guaranteed. More frighteningly still he was now compelled to believe in a universe that was virtually infinite in extent. Even if other worlds were to be inhabited his sense of lonely isolation could hardly be diminished.

There was another kind of discontinuity marked by the Scientific Revolution. Copernicus' idea of circular movements for the heavenly bodies was in due time displaced by Kepler's recognition that they actually moved in ellipses. Upon this foundation Isaac Newton established his whole celestial system, demonstrating that the self-same law (the inverse square law of gravitational attraction) applied to a falling apple and to a heavenly body swinging through the depths of space. This was the Newtonian universe: immense and boundless; rational, comprehensible and above all mechanistic. It was a far cry from the universe of even a generation earlier which had been envisaged as something much more akin to an *organism* than a *mechanism*. Even Copernicus himself displayed traces of the old tradition with his almost pantheistic reverence for the sun which ruled over his tribe of planets. This mechanistic picture of the universe led to Nature losing its capital N and at the same time its pretensions to divinity. Part of the preoccupation of the Enlightenment was to discover in what ways God might be related to the universe if He was no longer part of it.

The Scientific Revolution was one of the few very great watersheds in Western thought, comparable with the rise of Christianity, the Reformation and the Renaissance (to all of which it was closely related). From the Enlightenment emerged the first stirrings of what was later, and contentiously, to be called the Industrial Revolution. Beginning in Britain and spreading over Europe and America, the complex series of social changes due to new technology and capitalist enterprise has often been hailed as one of the triumphs of science itself. Yet it would be just as misconceived to depict the Industrial Revolution as a direct consequence of scientific advance as it would be to regard the immediate effects of the Scientific Revolution as technological progress. What happened in the period between the Scientific and Industrial Revolutions, therefore, is of great importance for our understanding of both intellectual and social changes that lie at the root of modern western civilisation.[2]

2.2 'LAWS SUBLIMELY SIMPLE'

These words represent a common eighteenth-century evaluation of the greatest achievement of the previous century's greatest intellect, of whom Thomson wrote:[3]

> Newton, pure Intelligence, whom God
> To Mortals lent, to trace his boundless Works
> From laws sublimely simple

However to see the science of the eighteenth century simply in terms of a logical – almost inevitable – progression from that of Isaac Newton is to invite well-deserved ridicule, for it enshrines an outmoded and whiggish view of history and ignores the great volume of recent research that has in some degree demythologised the story of Newton's later influence. Nevertheless, there is sufficient substance in the myth to justify a brief survey of 'Newtonianism' in the eighteenth century; but it will be necessary to trace the course of events owing little directly to Isaac Newton and (in a subsequent chapter) to introduce systems that were canvassed as alternatives to Newtonianism. Attention to Newtonianism is justified by the abundance of testimonies to its importance both by adherents and adversaries alike within the Enlightenment. None could have been more fulsome in his praise of Isaac Newton than Pope in his oft quoted couplet:[4]

> Nature, and Nature's Laws, lay hid in Night;
> God said, *Let Newton be!* and All was *Light*.

Yet even he presumably felt a need to remind his audience of Newton's humanity:[5]

> Superior Beings, when of late they saw
> A mortal Man unfold all Nature's Law,
> Admir'd such Wisdom in an earthly shape,
> And show'd a NEWTON, as we show an Ape.

Newtonianism is a complex term and the cluster of ideas associated with it is far from straightforward.[6] It may be helpful to indicate some of its elements as follows, though most of these are inter-related:

(a) Gravitational attraction between material bodies, acting at a distance without material intervention (e.g. across a void);
(b) In accordance with this, an inverse square law (i.e. the force is inversely proportional to the square of the distance between the interacting bodies).
(c) Validity of the experimental method.
(d) Concept of the uniformity of nature, with the same laws applying over the whole visible material universe.
(e) A conception of the universe as being maintained as well as created by God.

This is rather an odd list in the sense that the first two items relate to 'laws of nature', the third one to a means of becoming acquainted with such laws, and the last two to metaphysical assumptions about reality. Yet the very nature of a 'law' is implicit in (d) and the conception of action at a distance (a) was for Newton a direct consequence of his theological presupposition (e). Most discussions of Newtonianism in the eighteenth century, especially in England, showed an awareness of the composite nature of the subject. It is the contention of the next two chapters that important social consequences flowed from every one of these five propositions.

When Sir Isaac Newton died in 1727, his reputation was at its zenith. A state funeral at Westminster Abbey, accompanied by panegyric of the most effusive kind, would have led one to suppose that scientific developments of the greatest portent were just round the corner. The millennialist hopes of the previous decades – even of Newton himself – seemed about to be fulfilled. In the event things turned out rather differently, and science experienced the novel sensation of criticism and disparagement. Some of the reasons will emerge in the next chapters, but some at least related to the curious and composite nature of Newtonianism itself. As we have seen, it enshrined both a methodology and a generalisation. What needs now to be added is that the methodology was itself complex. In addition to the empirical method, advocating observation, experiment, and standing in the well-established Baconian tradition, there was also the method of mathematical reasoning.

Now it so happens that, for most men, the greatest achievement of Newton (the inverse square law of gravitation) represented the apex of a mathematical approach to nature which they themselves found incomprehensible. On the other hand, the empirical approach, which they could readily grasp, led to more modest advances: new insights into the nature of light, suggestive hints for the progress of chemistry and more speculative ideas on physical forces in the universe. What they did not realise was the enormous resources in time and energy demanded by the empirical method before it could yield tangibly beneficial results. The Laputan gentleman who had spent eight years 'upon a project for extracting sunbeams out of cucumbers' was Jonathan Swift's caricature of an academic practising Newtonian empiricism.[7] And of course fewer still were able to envisage any useful result to follow from the laws of gravitational attraction. Nevertheless it was the breathtaking magnitude of Newton's cosmology that overawed the minds of his compatriots. Thus in 1719 a Cornish naturalist Walter Moyle wrote of the Royal Society:[8]

> I find that there is no room in Gresham College for Natural History: Mathematics have engrossed all; and one would think the Gentlemen of that Society had forgot that the chief end of their Institution was the advancement of natural knowledge.

This domination by mathematics was, however, to be short-lived in England. Not only were the detailed mathematical arguments of Newton quite incomprehensible for the ordinary Fellow of the Royal Society, but also the *Principia* itself was not available in English until 1729. In England dynamical studies were further inhibited for another reason as the century wore on. The mathematical technique called the calculus had been developed by Newton but his approach was primarily geometrical and not best suited for some of the problems in the century that followed. On the other hand the differential calculus of Leibniz was being rapidly developed on the continent, but was anathema to most mathematicians in England because of bitter recriminations which had followed a priority dispute between Leibniz and Newton. In fact each had clung to his calculus independently but this weighed little with the defenders of Newton who regarded the issue as involving national honour. To make matters worse, in 1710 Leibniz had quite justifiably suggested to Newton that he had conceived a universe so imperfect that God had continually to tamper with it. These factors, combined with a dearth of natural mathematical genius in England during the Enlightenment, combined to put that country at a great disadvantage in terms of development of Newtonian dynamics. Thus it was on the continent, rather than in Newton's England, that the development of theoretical physics chiefly took place. Englishmen who could scarcely understand a word that Newton had written nevertheless felt bound to defend him as a national hero.

In France the reigning cosmology at the beginning of the eighteenth century was that of Descartes: a mechanical universe, it is true, but one filled with swirling matter whose vortices constituted planetary motion. Despite the total lack of evidence for such Cartesian vortices they were still being championed by the aged Fontenelle as late as 1752. By then he was almost a solitary voice, and the more strident tones of the satirical writer Voltaire had been arguing the case for Newton nearly twenty years before. With the help of the Leibnizian calculus, the growth of dynamical astronomy proceeded apace in both France and Germany and, by the end of the century, Laplace could produce his *Système du Monde* in which the Newtonian dream of a physically unified cosmos was fulfilled in all respects save one: Laplace was so successful that divine intervention as an explanation for anomalies was no longer required. There were no anomalies.

Meanwhile that other strand of Newtonianism was being woven into the fabric of science in England and on the continent. Experimental philosophy had long flourished in Calvinist Holland, and Newtonian empiricism found some of its early and most influential converts there, amongst them W. J. 'sGravesande, H. Boerhaave and P. van Musschenbroek. Their writings, as Rupert Hall has pointed out, 'played a large part both in promoting Newtonian (and more generally English) empirical science on the continent, and in transforming mechanics from the most forbidding of mathematical sciences to a pleasant series of demonstrations. They helped also, however, to spread the confused view that the *Principia* and the theory of gravitation were no less founded upon empiricism than was *Opticks*'.[9]

The *Opticks*, first published in 1704, dealt not only with Newton's experiments in light. It also contained, especially in the second edition of 1717, a series of 'Quaeries' in which Newton allowed his imagination free rein and made pregnant suggestions for future developments in the sciences. Despite Moyle's complaint that 'mathematics have engrossed all' it soon became clear that science could flourish at least as well in the spirit of the *Opticks* as in the spirit of *Principia*.

Indeed, association with the name of Newton could add status and respectability to many other kinds of scientific endeavour. Nor was this always a device to 'legitimate' existing practice. There is strong evidence that the Newtonian concept of universal law gave men a genuinely new appreciation of the world around them, especially now that it had lost the crudity of some of its early 'mechanism'. So if dynamical astronomy was to flourish on the continent in the bright beams of Newtonian light, there was also a substantial growth of non-mathematical, or semi-mathematical, science in what might be called the Newtonian shadow. It is to those developments that we now turn.

2.3 SCIENCE IN THE SHADOW OF NEWTON[10]

Whilst the elaboration of the Newtonian universe and the development of physical dynamics constituted the most spectacular achievement of eighteenth-century science, other advances took place to which Newtonianism made a much less obvious contribution. Certainly Newton himself was long before his death (1727) a cult figure of universal stature. Not only had he accomplished unbelievable feats in physical science itself; he had also come to symbolise progress in comprehending all parts of the nature that was, paradoxically, both infinitely more complex than previously imagined yet now more open than ever before to careful examination. Because of this cultic significance of Isaac Newton it does not follow that all public acknowledgements of his inspiration should be taken at face value. It is at least possible that some dedications to Newton were attempts to legitimate as 'science' activities that might otherwise have seemed less prestigious.

However, there is little doubt that in a general way Newton was influential far beyond the areas of mathematical physics and astronomy. And in some quite specific ways certain of the non-quantitative sciences developed in a characteristically 'Newtonian' direction. Optimistic hopes that mercantile interests would be served by a wider growth of scientific knowledge doubtless played some part in the new investigations of chemistry, electricity and botany. More significantly, the social changes which led to a greater availability of leisure gave both opportunity and incentive for a far greater deployment of amateur talent in science than ever before. As with post-Newtonian dynamics, astronomy and optics, the more qualitative sciences owed much of their progress to an underlying ideology that was a deeper expression of social changes in Enlightenment Europe. And that was also true of the most mathematical of investigations.

In 1700 the known electrical phenomena,[11] as we might call them, were an ill-assorted collection of facts relating chiefly to effects obtained by rubbing amber, glass and other substances. Such cases of 'frictional electricity' were manifest by attraction or repulsion of light objects such as feathers, threads etc. in the neighbourhood of the rubbed object, crackling sounds and even the appearance of small sparks. Sir Thomas Browne, who first used the word 'electricity' in 1646, meant by it merely a quality like that of amber (Greek *elektrum*), 'the power to attract straws and light bodies'. Most people followed Gilbert's notion (1600) of electricity as some kind of humid effluvium, and this concept of electricity as a fluid persisted well into the nineteenth century (one still hears, occasionally, references to 'turning off the juice'!). Whether one or two fluids were involved was a matter of frequent and heated debate. Strangely, it may seem, the fluid was not considered as flowing until Stephen Gray, a Charterhouse pensioner, became the first to give an explicit description of electrical conduction (about 1730). Even then, however, it was the transference of charges of static electricity along lines of packthread suspended by silk loops. Devices which would produce a continuous current, transmittable along metal wires, had to wait until a series of accidental discoveries was made at the very end of the century (Galvani and Volta).

Enormous interest in electricity was stimulated by the work of the Englishman Francis Hauksbee. Having assisted Robert Boyle in the construction and use of his air pumps, he was well placed to pursue the study of electricity at reduced pressures. He devised a means of generating frictional electricity in exhausted receivers, and noted the spectacular glow of the electric discharge (extinguished when air was admitted). His work, reported in *Philosophical Transactions* from 1713 to 1715, aroused considerable interest in electricity, especially amongst those who had the means and the leisure to conduct their own experiments.

The search for the spectacular was only partially satisfied by such demonstrations. Little could be done until larger and more frequent sparks could be obtained than those from the laborious process of rubbing glass rods with pads of leather or wool. Despite Hauksbee's work that was the only technique commonly available until the 1740s. Then, apparently under the stimulus of sensationalism, several workers in Germany invented devices in which glass globes were mechanically rotated in contact with the human hand or (better) a cushion of leather held against them by springs. These 'electrical machines' soon appeared with variations, in England and France. They enabled a continuing series of fat electric sparks to be obtained at will. A direct outcome was the discovery that charges of static electricity could be stored – the first electric condenser. A clergyman of Pomerania, E. G. von Kleist, had accidentally stumbled upon this fact in 1745, and a more famous discovery was that of Pieter van Musschenbroek at Leyden in 1746, who cherished the optimistic hope of being able to store electricity by passing it into a bottle of water, and then sealing it. Having generated electricity with a machine he passed it into the water, holding the bottle with one hand. On accidentally touching the connecting wire with the other hand he experienced a powerful shock ('in a word, I thought it was all up with me'), for the electric charge was in truth not merely on the inside of the bottle but also on the outside. This combination of two conductors (in this case hand and water) separated by a non-conductor (glass) was thenceforth known as the Leyden jar. If one of the conductors was earthed, large electric charges could be induced in the other; the electricity could be 'stored'. Again, some of the early uses of this discovery appear rather frivolous. Adventurous souls would gain much satisfaction from the administration of a mild shock. Nor was the entertainment always for the benefit of the subject of the experiment. In 1741 the Abbé Nollet, in Paris, repeated Musschenbroek's experiment and discharged a Leyden jar in spectacular fashion through a chain of no less than 180 monks. Some hopeful operators even tried to estimate the rate of the passage of the electric charge from the speed of the involuntary but often vigorous reactions.

The use of human beings as subjects for experimental purposes in electrical studies was not new. Stephen Gray had suspended a Charterhouse pupil from the rafters by means of silk threads. When charged glass rods were applied to his feet, the unfortunate youth discovered that his face would attract particles of feathers etc. and was therefore also charged. From the 1750s electric discharges were much used in medical practice in London and elsewhere chiefly, it seems, for their stimulating effect. Man himself was now fast becoming part of the physical universe and his role as object rather than merely subject, of scientific experiments was beginning to raise the problems about the place of man in nature that so exercised the Victorian mind.

It was by means of a Leyden jar that the American journalist and politician Benjamin Franklin was able to confirm his suspicions that lightning was electrical in nature. In his famous kite experiment of 1752, on the banks of the Schuylkill in Pennsylvania, electricity from a thundercloud was conducted safely to earth and stored in a Leyden jar. His invention of the lightning conductor, 'upright rods of iron made sharp as a needle', predated this experiment by a few months and was one of the few outcomes of eighteenth-century electrical research to be of immediate practical use. A study of atmospheric electricity became the chief electrical preoccupation of the Fellows of the Royal Society, and they experienced some frustration through the unusual scarcity of thunderstorms in England in the 1750s. It was less likely to be 'religious scruples'[12] than the general state of science in England that delayed the establishment of lightning conductors on St Paul's Cathedral, London, until 1768. It was to the somewhat lethargic Royal Society that the Dean and Chapter turned for advice after St Bride's Church in the City had been seriously damaged by lightning. Moreover, it can hardly have been theology that inhibited the British government from seeking advice on protecting its gunpowder magazines at Purfleet, but it was only in 1772 that the Royal Society ratified Franklin's advice to use pointed lightning conductors to protect the establishment. A minority report (by Benjamin Wilson) suggested genuinely scientific doubts, for would not pointed rods attract the lightning, overheat the rod and so 'promote the mischief they were intended to prevent'?[13] Although the majority report was accepted and pointed conductors installed, five years later there was a lightning strike at Purfleet, though fortunately no explosion. This, and one or two similar incidents elsewhere, prompted the Royal Society to re-examine the question, but their Committee unanimously supported the previous viewpoint. Once again Wilson, not this time on the Committee, expressed himself vociferously against the conclusion, and wrote accordingly to the Board of Ordnance. By now, however, England was at war with the American Colonies and Benjamin Franklin was their European representative. The affair had entered the political arena, and became a classic, if minor, example of abortive political attempts to meddle with science. As Sir Henry Lyons wrote:[14]

> Those who advocated pointed conductors were identified with the supporters of the insurgent colonists. The King is said to have taken the side of Wilson in the dispute on political grounds; he had blunt conductors installed at the palace, and even endeavoured to make the Royal Society rescind its resolution in favour of pointed conductors. In an

> interview with Sir John Pringle at the end of 1777, His Majesty is said to have urged him to use his influence in supporting Mr Wilson. The President however replied: 'Sire, I cannot reverse the laws and operations of nature'.

There seems to be substance in the view that the incident led to disenchantment with the Royal Society at Court, to Pringle's retirement and to the choice of Joseph Banks as his successor.[15]

Despite the prominence that electrical science enjoyed during the Enlightenment it was in an altogether different state of organisation from the more directly Newtonian sciences. Yet the studies in static electricity, electric discharges, attraction and repulsion, induction, and atmospheric electricity (to name but a few of the important topics) had led to the proliferation of one- and two-fluid theories, and even to the position of extreme empiricism adopted by Joseph Priestley who in 1767 had become 'one of the first to deny themselves the pleasure of declaring what electricity is'.[16] Thus it was said of Franklin that 'he found electricity a curiosity and left it a science',[17] though that is not to imply that the achievement was his alone. Thomas Kuhn has observed:[18]

> Some time between 1740 and 1780, electricians were for the first time enabled to take the foundations of their field for granted. . . . As a group they achieved what had been gained by astronomers in antiquity and by students of motion in the Middle Ages, of physical optics in the late seventeenth century, and of historical geology in the early nineteenth. They had, that is, achieved a paradigm that proved able to guide the whole group's research.

This is a far cry from Newton's tentative query as to why the electric effluvium had no weight yet it is nowhere near the state of quantitative perfection in which he had left dynamical physics. It is interesting to note that Coulomb's direct proof of an inverse square law for electrical repulsion (1784) came exactly a century after Newton sat down to pen the first pages of *Principia* in which he announced the inverse square law for gravitational attraction.

If 'revolution' came late to the science of electricity, this is even more true of chemistry.[19] Herbert Butterfield has spoken of 'the postponed scientific revolution in chemistry',[20] by which he meant the radically new direction given to the subject by Lavoisier at the end of the eighteenth century. Lavoisier's researches cleared up in large measure the historic difficulty of defining and identifying the ultimate kinds of substance of which all materials are made – the chemical elements. A tentative attempt to arrive at a new view of chemical elements had been made in 1660 by Robert Boyle, and had gone some way to undermine traditional notions of the elements, including the Aristotelians' air, earth, fire and water. But a century and more was to elapse before the key to further progress was to be found. That key lay in the most familiar of all chemical processes. As the Swedish chemist J. J. Berzelius observed much later:[21]

> The theory of combustion (and accompanying phenomena) has always been the basis of chemical theory, and it is probable it will never cease to be.

The phenomenon of fire was widely associated with electricity in the years following Franklin's success in drawing 'electric fire' from the heavens. Even more important as a generalisation about combustion was the theory of phlogiston (with which indeed certain electrical explanations were sometimes amalgamated).[22] This theory, sometimes called the first great generalisation in chemistry, stated that during combustion something was given off into the atmosphere, as indeed often appears to be the case. This 'something' was phlogiston, the principle of inflammability. Originating in Germany with the work of J. J. Becher (1669) and his populariser G. E. Stahl (1703), the theory was being accepted in France by 1730 and by mid-century had become the basis of chemical orthodoxy. When a material (say zinc) burns the phlogistonists would say:

$$\text{Zinc} \longrightarrow \underset{\text{(into the air)}}{\text{phlogiston}} + \underset{\text{(solid residue)}}{\text{calx}}$$

All combustibles contained phlogiston by definition. Similarly the extraction of a metal from its ore would involve uptake of phlogiston (e.g. from charcoal), and most chemical reactions, even including respiration, could be viewed as a transfer of phlogiston. As a unifying principle in chemistry – a paradigm – it could hardly have been overthrown without a revolution, although when the chemical revolution was initiated by the French chemist A. L. Lavoisier it was about much more than combustion and phlogiston. Its central feature was a recognition that air was not elementary but had two main constituents, nitrogen and oxygen. The role of air in combustion is not then merely as a dumping-ground for phlogiston, but for its minor constituent, oxygen, to combine with the combustible to form an oxide. Thus the calcination of zinc would be re-expressed thus:

$$\text{Zinc} + \text{oxygen} \rightarrow \text{zinc oxide (calx)}$$

This oxygen was on Lavoisier's scheme the principle of acidity and, moreover, just one of a number of gaseous elements, as may be seen from the list given in Lavoisier's *Traité*, one of the seminal documents of the chemical revolution.

But between Boyle and Lavoisier one train of events had transformed the science and made Lavoisier's revolution possible. This was the rise of pneumatic chemistry, the chemistry of gases. Before 1700 the notion of gases as chemical individuals hardly existed. Very few had been examined and these were considered merely as variations of air. Then, in the early years of the eighteenth century, the Reverend Stephen Hales, Perpetual Curate of Teddington, took up a suggestion of Newton's and heated various bodies, measuring the quantities of gases ('airs') evolved. To do this he made substantial improvements to existing techniques for manipulating gases, and in fact collected many of them over water. Later on the Reverend Joseph Priestley, a dissenting minister, extended the technique to soluble gases by collecting them over mercury, in this way discovering ammonia and several other gases. He made numerous other technical improvements and actually discovered oxygen independently of Lavoisier (though did not draw the same epoch-making conclusions). Meanwhile the Hon. Henry Cavendish collected gases in bladders of known capacity and thus identified 'inflammable air' (hydrogen) as a chemical individual. From this it followed that water was not an element but a compound of hydrogen and oxygen. In Scotland Joseph Black examined 'fixed air' (carbon dioxide) and demonstrated its relationship to the mineral carbonates. All these discoveries made possible the chemical revolution of Lavoisier and go at least some way to explaining the 'postponement' that so troubled Butterfield. It is noteworthy that almost all the experimental work on gases preceding Lavoisier was performed in Britain.

It has been sometimes said that Lavoisier succeeded where others had failed because he introduced quantitative measurements into chemistry. To be sure, he weighed his metals before combustion and weighed calces afterwards, the increase in weight suggesting therefore that something had been added to them (oxygen). Unfortunately this claim is a gross oversimplification. An increase in weight on combustion had been noted at least since the time of Boyle, and workers in the chemical crafts had been weighing their assay samples with a considerable degree of sophistication since the middle ages. However, there is one sense in which Lavoisier did bring a new quantitative approach to chemistry. In 1768, the year of his admission to the French Académie, this young lawyer decided science was now to be his chief preoccupation and, to finance this ambition, he became one of the Farmers General, who collected indirect taxes for the Government and made substantial profits themselves. There is at least a congruence between the commercial balance sheets with which he was familiar as a taxman and the notion of a balancing chemical equation based on a recognition of the law of conservation of matter.

This law had not been explicitly stated before Lavoisier. Moreover he had an excellent quantitative technique with a very good balance obtained from the Inspector of Coinage, giving him an accuracy far greater than Boyle was able to enjoy. This method of scrupulously accounting for all the weighable material in a reaction was now applied to pure substances, and so gave a wholly new view of chemical change. It was a strange irony of fate that when Lavoisier was sent to the guillotine in the French Revolution it was on a trumped-up charge of adulterating tobacco – of fraudulently fudging his quantitative results.[23]

During the eighteenth century chemistry was striving for quantitative precision in another way as well, very different from that taken by Lavoisier. This was seen in the series of attempts to study the short-range forces between atoms, just as Newton had considered the long-range forces that caused gravitation. Newton himself had repeatedly alluded to such forces and had given birth to what one author has called 'the Newtonian dream'.[24] This was to unite chemistry by a mathematical treatment of its data. Several early workers in England committed themselves to speculate along these lines, and the work of Hales owed something to this tradition. By mid-century the main arena for such 'Newtonian' exercises shifted from Britain (or Holland) to France, with a recrudescence of the so-called 'affinity tables'. These were attempts, however crude, to give some numerical value to the avidity with which chemical substances combined with each other, their 'affinity'. In fact the parameters involved (weights of combining substances) are measures of something quite different and were to lead to later conclusions of great importance. But their occurrence is another broad indication of the belief that if chemistry was to be a branch of natural philosophy then quantification was essential.

A good deal of the chemical writing in the eighteenth century can be seen as reflecting its ambiguous status in that period. For this its hybrid ancestry was responsible.[25] The intention of the Newtonians, of course, was to bring it within the sphere of natural philosophy. A clear lineage may be traced back to the natural philosophy of Newton and of Boyle, and hence to alchemy itself. But there was also a separate tradition of chemistry as a useful art, performed by assayers, vitriol manufacturers, dyers, brewers, soap-boilers and many more. Short on theory, secretive and with as wide a range of abilities as of skills, these men were reluctant to communicate their knowledge, but nevertheless were laying the chemical foundations of the later Industrial Revolution. Their separation from natural philosophy was not quite complete, however. (Was not Newton himself Master of the Mint, and did not some of the leading manufacturers like Roebuck, Keir and Hutton study chemistry at Leyden or Edinburgh?) But the difference was sufficiently great for the French metallurgist

Gabriel Jars to write of the alum works at Whitby in 1765 that 'alum is, of all vitriols, the hardest to make, and the management of such an establishment needs a chemist experienced in large-scale operations'; however, in 1757 he had complained of the Schwemsel works in Saxony that 'there is no weighing of the water as in England'.[26] That is entirely typical of alum manufacture in the eighteenth century and of much else in manufacturing industry; chemical processes were carried out in a totally empirical way.

The third tradition in chemistry was pharmaceutical. Early in the eighteenth century drug manufacture in England was largely in the hands of the apothecaries, but gradually an independent trade developed so that, by 1774, the Society of Apothecaries restricted its livery to those in medical practice. A typical case was the Quaker apothecary Sylvanus Bevan (1691–1761) who in 1715 founded the famous Plough Court Pharmacy in London – a predecessor of the firm of Allen and Hanbury.[27] Its distinctive emphasis was the production of chemical medicines, Bevan himself cultivating medicinal plants in his own garden.[28] The eighteenth century saw the introduction of numerous chemical substances into medicine, most notably perhaps digitalis (Withering, 1785) and arsenic (Thomas Fowler, 1786). In 1747 Robert James, an associate of Samuel Johnson, patented his fever powder. His specification is remarkable both for its vagueness and lack of any quantities; but it does indicate something of the techniques available at the time:[29]

> Take antimony, calcine it with a continual protracted heat in a flat unglazed earthen vessel, adding to it from time to time a sufficient quantity of any animal oil and salt well dephlegmated; then boil it in melted nitre for a considerable time, and separate the powder from the nitre by dissolving it in water.

This powder was widely used, and the second half of the century saw a marked increase in the number of patent medicines available in England, as well as an increasing range of chemical substances being used. Some of these were vegetable in origin, while others were obtained ultimately from minerals. In either case large collections of available sources were required, and in that sense chemistry became dependent upon the skills of the natural history collector. Partly for this reason chemistry before 1840 has been well described as 'the natural history of the inanimate',[30] but there is no doubt that for most of the century it hovered uncertainly on the borderlines of natural history, natural philosophy and useful art.

Simply in terms of the number of adherents, natural history itself was the most important branch of science in the eighteenth century. It made the least demands intellectually, offered the most immediate satisfactions in terms of acquisition and physical exercise, and was in close harmony with the reigning ideologies of the Enlightenment. The collecting and arranging of specimens may not seem particularly well up in the hierarchy of scientific skills. Raven has called it 'an adolescent phase',[31] but that is also an appropriate description of the state of much biological and geological science in the eighteenth century. There was indeed a sense in which biology, especially, could not hope to make further progress until chemistry and physics were more advanced. To quote Raven again:[32]

> Anyone who studies, for example, the attempts of seventeenth-century physiologists to explain respiration in terms of a fiery element in the chest, or to investigate nutrition and digestion, or to appreciate the relationship of a plant to the soil will realise that biological studies had almost reached a point at which they must mark time till sound foundations had been provided for further advance.

So eighteenth-century botany had to be content largely with naming and describing plants. There was plenty to be done, with new specimens constantly entering England, France, and Holland with travellers from overseas. The super-abundance of exotic new species, however, merely intensified the discomfort of the botanists. How could the immense sum of knowledge now available be codified and ordered in a systematic way? The problem was new only in its scale. In 1704 the final volume of John Ray's *Historia Generalis Plantarum* completed his account of over 18,500 plants, arranged in 125 sections. He also wrote on fishes, quadrupeds, serpents and insects. But he continued to use cumbersome biological names and, except in England, was not widely recognised. One of the great pioneers of natural history in France, was Buffon the well-to-do *Intendant* of the Jardin du Roi (appointed 1739). He was required to catalogue the Royal collection. Thus began a great *Histoire Naturelle* (44 volumes), intended to encompass the whole of the subject, and founded on his own classification. All creative things were part of a great chain of being[33] (no new concept this), from air and water and stones through plants, insects, shellfish, reptiles, other fish, birds, quadrupeds up to man himself. This continuous chain he divided in what he deemed a 'natural' manner, on the basis of habitat and even usefulness. His system was widely supported in France until well into the nineteenth century.

The greatest advance in classification came from the Swedish botanist Carl Linnaeus. A passionate classifier, he produced an alternative system which, for plants at least, was based on their newly discovered sexuality. This first appeared in his *Systema Naturae* (1735–58), in which he was able to place every known plant and animal, first in a *class* then in an *order*, then a *genus* and finally a *species*. In England and France the merits of Ray and Buffon, respectively, were vigorously canvassed in the face of this northern challenge to their authority. By about 1760 the Linnaean System was making many converts, perhaps above all because it happened to be combined with a proposal for binomial nomenclature. Living creatures were named by reference to their genus and species – two words only. This way the task of reference and discussion was greatly facilitated, and despite cries from Buffon of 'artificiality', the Linnaean scheme (in a modified form) has become the foundation of modern biological classification. Its originator has been immortalised not only by his system but also by his splendid collection of specimens, now in London and the property of the Society which bears his name. It is perhaps no coincidence that natural history in England took an upturn just as the Linnaean System was being widely accepted, although it may not be unconnected with the accession of George III who, together with his wife and mother, was known to have a keen interest in matters botanical.

3. Nature in the Enlightenment

3.1 NATURE AND IDEOLOGY

DESPITE the advances in science recorded in Chapter 2 – and partly because of them – it is difficult to avoid the conclusion that science was generally in poor shape during most of the eighteenth century. This was reflected in its institutional life (Chapter 5), in the quality and number of published works, and in the diminished status acknowledged to its practitioners by some of their contemporaries. Thus the English physician David Hartley (1705–57), a founder of the associationist school of psychology, could observe that 'nothing can easily exceed the vain-glory, self-conceit, arrogance, emulation, and envy, that are found in the eminent professors of the sciencies'.[1] He was far from being alone in attributing 'ill passions' to men of science.

In searching for the causes of this decline in achievement and reputation commentators have frequently found refuge in the sterility of Newtonianism itself, a consequence of both the magnitude and completeness of Newton's own work in natural philosophy. But Newtonianism meant also a commitment to the experimental method, and in no sense could that be said to have reached its summit in the eighteenth century. So further explanations must be sought; thus J. D. Bernal has attributed the English decline as being 'to a far larger extent . . . due to social and economic factors'.[2]

Unfortunately he seems to identify these as simply replacement of the 'gentlemen merchants' who founded the Royal Society by the wealthier Whig aristocracy, who were more interested in speculative finance than in science. This was clearly illustrated by the declining fortunes of the Royal Society. It may be doubted, however, whether a crude economic model of this kind is adequate. It purports to be a semi-quantitative analysis ('to a far larger extent') though without evidence; it has nothing to say about the parallel down-turn of science outside England, though Bernal admits that this occurred 'to a lesser degree'. More seriously it underrates the importance to science of *techniques*, whose emergence may have nothing to do with 'economic factors' and yet were critically important for chemistry, electricity and other sciences.[3] Towards the end of the century, London was recovering its scientific reputation and becoming a Mecca for continental men of science. The Dutch physicist M. van Marum (1750–1837) was one of a long line of scientific pilgrims seeking, amongst other things, 'the finest instrument market in the world'.[4] Contemporaries at least saw techniques as being very important for the development of their science. Again, it is important to remember that science needs *systems*, whether the Linnaean classification in biology, the mid-century paradigm for static electricity, or the anti-phlogiston system for chemistry. And science in the eighteenth century needed above all an *effective organisation*.

It is rather hard to see how these considerations could be regarded as less significant than the nature of the ruling class. One may too easily assume that patronage played a critical role in scientific progress in the Enlightenment. To be sure, the quickening pace of scientific enquiry under an enlightened patronage may bring to birth new systems rather more quickly than otherwise; but it may also give rise to systems that are less than helpful and which, in retrospect, are seen to impede the progress of science. It may be argued that this was true of the theory of phlogiston, introduced by J. J. Becher and G. E. Stahl, who enjoyed the protection of, respectively, the Elector of Bavaria and Frederick I of Prussia. Moreover in eighteenth-century England plenty of discoveries in astronomy, chemistry and (especially) natural history were made by those remote from the ruling class. For evidence of this one needs to go beyond the pages of *Philosophical Transactions*, and to examine patents, papers in some of the short-lived naturalist periodicals, industrial practice, letters, diaries and, as the century proceeded, books and pamphlets. What does become clear at the end of the eighteenth century is that science in its changing modes of expression and practice reflected underlying changes in the social ideologies of the period. Most obviously these will be manifest in the organisation and control of scientific experiment, and in the dissemination of scientific knowledge. It may also be argued that they would have some effect on the actual science that is performed and on its interpretation. These ideological changes must now be examined, affecting in every case man's view of one or other aspects of nature.

3.2 NATURE AND MAN

Nature in the Enlightenment was contemplated with a new interest and fascination by mankind, and in particular by men and women relating to it as individuals. For this was the great age of individualism, a fruit, no doubt, of the puritans' teaching of the infinite worth of each individual person, and now sanctioned by the realisation that individuals could wield considerable economic power given favourable situations. The tendency was reflected in the philosophy of Locke, in the novels of Richardson and Fielding, and in the countless acts of individual enterprise with which tradesmen and small merchants pursued economic goals which their fathers would have deemed improper or impossible (or both).

Nowhere is an individual's enthralment to the wonders of nature more sensitively described than in *The Seasons* by the poet James Thomson. First appearing in 1726, it reveals a depth of feeling combined with a close attention to detail:[5]

> Clear Frost succeeds, and thro' the blew Serene
> For Sight too fine, th' Ætherial Nitre flies,
> To bake the Glebe, and bind the slip'ry Flood.
> This of the wintry Season is the Prime;

Pure are the Days, and lustrous are the Nights,
Brighten'd with starry Worlds, till then unseen.
Mean while, the Orient, darkly red, breathes forth
An Icy Gale, that, in its mid Career,
Arrests the bickering Stream. The nightly Sky,
And all her glowing Constellations pour
Their rigid Influence down: It freezes on
Till Morn, late-rising, o'er the drooping World,
Lifts her pale Eye, unjoyous: then appears
The various Labour of the silent Night,
The pendant Isicle, The Frost-Work fair,
Where thousand Figures rise, the crusted Snow.
Tho' white, made whiter, by the fining North.

From now on the boundless works of God should be studied with the care and reverence demanded of an act of worship. Thomson's poem had immense popularity in the eighteenth century. Many years later a copy of *The Seasons* was a prized possession of the young Humphry Davy.[6] The transformation that Thomson reflects is unmistakable. Nature is now viewed in a colder, clearer light than before, and is seen as an object of study and analysis, and is worthy of minute description. A similar passion for nature appears in the poems of Thomas Gray (1716–71), who had annotated his own copy of Linnaeus' *Systema Naturae*, and whose famous *Elegy* begins:[7]

The curfew tolls the knell of parting day,
The lowing herd winds slowly o'er the lea,
The plowman homeward plods his weary way,
And leaves the world to darkness and to me.

Now fades the glimmering landscape on the sight,
And all the air a solemn stillness holds,
Save where the beetle wheels his droning flight,
And drowsy tinklings lull the distant folds,

Save that from yonder ivy-mantled tow'r
The moping owl does to the moon complain
Of such as, wand'ring near her secret bow'r,
Molest her ancient solitary reign.

Thomson is sometimes criticised for an over-prosaic concern for fine scientific detail, but in Gray this was replaced with a warm sensitivity to the wonders he recites. Something of the new appreciation of natural detail may be also seen in the cruel realism of Hogarth's cartoons, in the vivid depictions of horses by George Stubbs (1724–1806), and (to some extent) in the landscapes of Thomas Gainsborough (1727–88). Its supreme literary appearance must surely have been in the *Natural History of Selborne*, written in 1789 by Gilbert White (1720–93), a Hampshire parson whose idyllic pictures of rural life in all its vivid detail formed a quintessentially English contrast to ominous events taking place across the Channel.

Similar attitudes to nature were to be found in other countries. In mid-century France the *philosophes* were urging a vast detailed survey of all nature, comprehending even the phenomena of society. Their emphasis was if anything more Baconian than that in Britain: fact-collecting on a truly cosmic scale. Some of the contributors to the *Encyclopédie*, as Buffon and Diderot, regarded mathematics with suspicion if not contempt. Diderot reckoned that 'within a century there will not be three great geometers in Europe'.[8]

Ironically it was in France that dynamical astronomy was to be promoted with such outstanding success. The work of d'Alembert, Laplace and their mathematical colleagues amply confirms the strength of both kinds of Newtonianism in France as well as in England.

An Englishman would be interested in natural phenomena chiefly when they were on a fairly small scale. William Derham could gaze with infinite delight upon a tiny insect, reflecting upon 'the surprising minuteness, art and curiosity of the joynts, the muscles, the tendons, the nerves, necessary to perform all the motions of the legs, the wings, and every other part'. But in the case of large-scale geological features like hills and valleys he sympathised with the 'peevish weary traveller' for whom 'they seem incommodious and troublesome'.[9] In this respect Derham was taking more than a cue from *Telluris theoria Sacra* (1681 and 1689) by Thomas Burnet (1635–1715), a Cambridge cleric later to become Chaplain to the King. During a continental tour in 1671 Burnet had crossed both Alps and Appenines. This is what he thought of them:[10]

> The sight of those wild, vast and indigested heaps of Stones and Earth did so deeply stir my fancy, that I was not easie until I could give myself some tolerable account how that confusion came in Nature.

His explanation (they were 'the ruins of a broken world', debris left by Noah's flood) attracted much attention. The emotion which called it forth in the first place became routinely expected during the Enlightenment. Partly, of course, this was a natural response to the hazards of travelling through territory that was inherently dangerous, unmapped with any accuracy, and a likely haunt of thieves and other undesirables. As Thomas Gray records of Seathwaite:[11]

> All farther access is here barr'd to prying Mortals, only there is a little path winding over the Fells, & for some weeks in the year passable to the Dales-men; but the Mountains know well, that these innocent people will not reveal the mysteries of their ancient kingdom, the reign of Chaos & old Night. Only I learn'd, that this dreadful road dividing again leads one branch to *Ravenglas*, & the other to *Hawkshead*.

This was not all. The mountains failed in any obvious way to demonstrate the feature of nature so assiduously sought in the early eighteenth century: design. The views of Burnet lingered on into the Enlightenment:[12]

> [Mountains] have neither form nor beauty, shape, nor order no more than Clouds in the Air. Then how barren, how destitute, how naked are they? How they stand neglected by Nature?

Lacking in symmetry and proportion, they often offended by their sheer Uselessness. So at least were the opinions of Daniel Defoe, speaking of the hills of the Lake District:[13]

> Nor were these hills high and formidable only, but they had a kind of unhospitable terror in them. Here were . . . no lead mines and veins of rich oar as in the Peak, no coal pits,

as in the hills above Halifax, but all barren and wild – and no use or advantage either to man or beast.

Insects, fishes, and plants were comprehensible within a teleological conception of nature, particularly with man's well-being as their immediate end. The fact that nature on the grand scale failed to fit into this conception, and further diminished man by virtue of its sheer size, may have been a principal reason for (in Marjorie Hope Nicholson's phrase) 'mountain gloom' at the very end of the seventeenth century. And yet, as she pointed out, it was not long before 'mountains had ceased to be monstrosities and had become an integral part of varied and diversified nature'.[14] It required an effort to see beauty and usefulness in them, but Derham was able to challenge Burnet on his own grounds and to argue that 'mountains are so far from being a Blunder of Chance, a Work without Design, that they are a noble, useful, yea a necessary part of our Globe'[15]. Others welcomed a challenge, for 'imagination loved to be filled with an object or to grasp anything that is too big for its capacity'.[16] Gradually, the fascination with 'these stupendous works of nature' became tinged with pleasure as well as horror. This may be seen to perfection in Gray's lyrical description of Derwent Water at dusk, published in his *Tour of the English Lake District* (1769). Gray anticipated that new social phenomenon, the tourist. When Thomas West produced his *Guide to the Lakes* (1778) he prescribed a series of statutory viewpoints which had to be 'collected' by the tourist just as his father had collected fossils or butterflies. Gilpin's *Picturesque Tours* (1786) even advised examining the view through plano-complex mirrors.

Recognition of scenery (the word first appeared in 1784) represents for mankind a profound change in attitude towards nature. According to Paul Shepard[17] this occurs when man can regard himself as separate from nature and can aspire to a detached scientific vision. If this is so, then tourism is a direct social consequence of change of scientific attitudes (although as we have seen those attitudes had become more detached much earlier with respect to nature on the smaller scale).[18] But of course other more obviously social considerations are also relevant, not least the greater ease of travel and the mobility of population. In response to the danger of further Jacobite revolts, General Wade commenced an extensive road building campaign in the Highlands in 1725, creating nearly 250 miles of military roads within the next twelve years. The atrocious conditions of these and later roads in the second half of the century did not deter travellers of many kinds, including Boswell and Johnson in 1773.[19] The English campaigns of 1745 against the Highland rebels in any case introduced at least some southern Volunteers to the magnificent scenery of Northern Britain.

With all these changes man was becoming familiar with nature, which he could now contemplate with some detachment. It took some time for scientific appreciation to degenerate into stark utilitarianism, but the origins of the transformation were most clearly seen in the early eighteenth century. The mania for collecting that swept Europe at about this time in the frantic hunt for specimens of botany, entomology, palaeontology, mineralogy and other branches of natural history, undoubtedly led in due course to the development of those subjects as systematic sciences. How these activities were organised will be discussed in Chapter 5.

For nature 'in the large', the effect may be seen in the 'naturalistic' gardens of the Enlightenment, particularly in England. Unlike the mathematical and formal arrangements of the seventeenth century, the new gardens of Capability Brown and his colleagues gave to the owners of great country houses like longleat, Ashridge, Castle Ashby, Chatsworth and Blenheim, a prospect of great beauty and simplicity which looked natural but was in fact contrived. If it was true that nature was most to be admired when she conformed to the aesthetic standards of the day, so equally art was at its highest if it could produce tolerable imitations of nature. As a frontispiece to the 'designs by Mr R. Bentley for six poems by Mr T. Gray' it was appropriate to have a picture of art as 'nature's ape'.

As for mountains, the new relaxed approach of the late Enlightenment is well conveyed by the title of a map of the Lake District appearing in 1760 'A New Map of the Counties of Cumberland and Westmoreland divided into their respective Wards; From the best surveys and intelligences; Illustrated with historical extracts relative to Natural History,

Produce, Trade and Manufactures; Shewing also the Rectories and Vicarages; with various other improvements'.[20] As E. W. Hodge observes 'this is far from the simple dignity of the seventeenth century with its '*Cumbria Comitatus Vulgo* Cumberland'.[21] Nature was now to be seen, measured, classified, understood, used and improved. Nearly half a century was to elapse before those same mountains were to regain their awe and wonder in the romantic appreciation of William Wordsworth. By then nature and man had entered into a profoundly different phase of their relationship.

Meanwhile it is only necessary to add that these activities displayed in an obvious and even arrogant form the confidence possessed by individuals who asserted their rights to manipulate nature in a spirit of competition with their rivals. Amongst the early industrial pioneers like Watt, Wedgwood, Roebuck, Boulton and others there is a clear-eyed intention to control nature not merely for personal satisfaction but also for private financial gain. Their papers are searched in vain for declamatory rhetoric to justify these attempts; for them 'nature' and 'science' were less heavily charged with philosophical meaning than for the gentry and the intellectuals. Their freedom with nature may or may not have had much to do with what we should call science in any direct and demonstrable fashion; it had everything to do with the changed perceptions of man's individual relationship with nature that are directly traceable to the science of the previous generations, the science of Isaac Newton.

3.3 NATURE AND GOD

The idea that a study of nature should be kept quite separate from all matters of theology would have seemed extremely odd to most of the early Fellows of the Royal Society. Their counterparts in the Paris Académie would have been rather less surprised. From Newton to Darwin the 'holy alliance between science and religion' has been (as Basil Willey put it) 'that typically English phenomenon'.[22] It is no exaggeration to say that the effects of this alliance have been directly felt right up to the present day.

We are now concerned, however, to see how a changed understanding of the relationship between God and his creation had important consequences for both science and society in the Enlightenment.

As the eighteenth century opened, polite society in London was far from unaware of the achievements of natural philosophy. Apart from discussions in the Royal Society, and gossip in the coffee houses, a rather curious feature of metropolitan life had now become an established means of communicating science: eight sermons a year delivered to crowded congregations in one or other of the City churches. These were the famous Boyle Lectures, founded under the will of the Honourable Robert Boyle (1627–91) 'for proving the Christian religion against notorious infidels, viz atheists, theists,[23] pagans, Jews, and Mahometans, not descending lower to any controversies that are among Christians themselves'.[24] Although they were endowed by one of the great scientific figures of the seventeenth century, their connection with natural philosophy went much further. Much of the apologetic of the Boyle Lectures was in fact drawn from the discoveries of science. The very first Lecture by Richard Bentley[25] in 1692 concluded with three sermons specifically devoted to Newtonian natural philosophy. Before publishing his sermons the preacher had consulted with Newton himself and received the famous reply:[26]

> when I wrote my treatise about our system, I had an eye upon such principles as might work with considering men for the belief of a deity, and nothing can rejoice me more than to find it useful for that purpose.

In fact most of the Boyle Lectures were published, sometimes in translation, and became major vehicles for the propagation of science, and especially Newtonian natural philosophy, for much of the eighteenth century. The selection of Lecturers, and their choice and treatment of subject matter, suggests a formidable alliance between Newtonianism and the Latitudinarian wing of the Church of England, who saw in the new science a powerful weapon for routing the enemies of the Restoration Church and re-establishing belief in a God who designed, created and upheld the existing natural order. Whether their ambition extended beyond the purely theological will be discussed in the next chapter.

The Boyle Lectures marked a major development in the scope and scale of what has been called 'natural theology'. This phrase has a very specific meaning and implies a process of deduction about the being and attributes of God from the study of the character of his works in nature. In other words, it serves an essentially apologetic function, a sustained 'argument from design'. Despite Darwin, natural theology is by no means dead even today. Its finest flowering in the work of William Paley owed much to the writings of the early eighteenth century, though the Boyle Lectures were not the first published works of natural theology. It may be seen in the classic study by John Ray of 1691, *The Wisdom of God Manifested in the Works of Creation*,[27] and before that in Bishop Wilkins' publication of *The Principles and Duties of a Natural Religion* (1675), and above all in the writings of Robert Boyle himself.

Boyle had embraced the mechanical philosophy of the seventeenth century by which magic and subtle 'influences' were banned, all the phenomena of nature attributed simply to matter in motion, and an absolute distinction made between matter and spirit. The universe is 'nothing but' a vast machine and its workings are, in principle, understandable by human reason. Thus as he said in a famous passage, the world is 'like a rare clock, such as may be that at Strasbourg, where all things are so skilfully contrived, that the engine being once set a-moving, all things proceed according to the artificer's first design.'[28] Since then many writers, including Paley, have been captivated by the analogy. Just as a clock or watch implies a designer, so does the infinitely more complicated universe of nature. (The analogy long predates Boyle, of course, but he gave it an unusual prominence.)

In France the force of this argument was considerably lessened by the opposition of Descartes, a man who had paradoxically done more than any other to establish the mechanical philosophy. The nub of the disagreement between Boyle and Descartes – and to some extent between England and France – lay in the question whether it was possible for mortal man to understand God's purposes in design, that is, to discover final causes in nature. Descartes had argued that the divine purposes could only be known through revelation, though he admitted man's natural ability to discern the existence of God.

Boyle went much further and concluded that man was under a positive obligation to discover, if he could, God's purposes in nature. In this way he also gained extra sanction for his own love of experiments and scientific enquiry. In England Cartesianism received a crippling blow from quite another source, and much of Descartes' cosmology (with its space filled by whirling vortices) was shown to be inconsistent with the Newtonian system where gravity could act across great distances through empty space. It is not surprising that on these matters French and English thought diverged considerably. In France the mechanical universe became increasingly associated with materialism in the writings of the *philosophes*. As Butterfield has observed:[29]

> It was not the new discoveries of science in that epoch but, rather, the French *philosophe* movement that decided the next turn in the story and determined the course Western civilization was to take. The discoveries of seventeenth century science were translated into a new outlook and a new world-view, not by scientists themselves, but by the heirs and successors of Fontenelle.

That world-view was the materialist philosophy inherent in much of the *Encyclopédie* and expounded by P.-H. Dietrich, Baron d'Holbach (1723–89) in his *Système de la Natur* (1770). All was reduced to atoms and movement, even life itself; gone were 'soul' and 'spirit', and all that could be said of God could as well be said of 'nature', the only ultimate reality. No wonder d'Holbach found his prime targets in the Boyle Lectures.

In England the growth of natural theology was seen as a necessary counter-blast to such radical and dangerous doctrines. Some of these were varieties of atheism, which amongst other detestable features, could be supposed to have connections with seditious enterprises.

Natural theology was linked with orthodox Christianity and

also with deism. This was belief in a God who created the universe once and for all and thereafter left it to its own devices. The clockmaker, having made his timepiece, wound it up and let it go without further attention. Boyle, who never believed such things, was faced with a serious dilemma: the more completely the universe resembled a machine the more eloquently did it testify to God's design but the less cogently did it speak His providential care.

It is a measure of his dedication to the design argument that he was prepared to live with such a defect in his celebrated analogy.

Deism received its first major exposition in *Christianity not Mysterious* (1696) of John Toland (1670–1722), its chief manifesto ('the deists' bible') being *Christianity as old as creation* (1730) by Matthew Tindal (1655–1733). This was essentially a religion of reason in which revelation (i.e. scripture) was downgraded to a point where the Christological content of Christianity was reduced almost into a minimum. Thus it had affinities with other contemporary movements such as Arianism and Socinianism, in both of which the divinity of Christ was effectively denied. It is sometimes said that deism in England was suppressed by 1730, but its effects were felt long after in, for example, the uniformitarian geology of James Hutton. Its imprint may be discerned in the doctrinally emasculated view of religion confided to his diary by the young Humphry Davy:[30]

> The simple and fundamental truths of the Christian religion are perfectly intelligible and should be made the basis of faith; they will bear the test of reason; these will be firm and immutable amidst the eternal revolutions of opinion; these will exist as long as man exists. By these I mean the unity of God, the necessity of morality and the future state of retribution founded on the resurrection.

The crucial problem emphasised, if not invented, by natural theology was thus the problem of providence. Does God, or does he not, intervene in the universe that he has once made? To this question the eighteenth century gave four different answers. Each of them had considerable social and political significance.

1. God never intervenes in the universe. This view of God as an absentee landlord was of course the characteristic note of deism. The more strongly law, regularity, and uniformity were emphasised the more effective this view would become.

2. God intervenes only on rare occasions. This view was taken by Boyle[31] in connection with the biblical miracles. They were highly exceptional, as were such interventions in modern times. On the other hand Samuel Clarke, in his Boyle Lecture of 1705, tended to believe that some strange phenomena might imply 'an extraordinary interposition either of God himself . . . or at least of some intelligent agent far superior to man'.[32] It was Newton himself who gave the notion of an occasional adjustment greater credibility since gravitational principles alone could not explain all the observed phenomena in the solar system. This early case of a 'God-in-the-gaps' explanation was to cause much trouble later on, but that is another story.

3. God exerts his control primarily through the will of men. Thus, Samuel Clarke in expounding the Lord's Prayer distinguishes three meanings for the phrase 'the kingdom of God'. If it means 'the kingdom of nature' there is little point in praying 'thy kingdom come', 'because it is at all times actually, at all times necessarily . . . present'. It can also be a 'kingdom of glory' in the future, and a 'kingdom of grace', where God exerts dominion *over the wills and actions of free agents*'.[33] It is here that God is able freely to act, subject only to the compliance of his creatures.

4. God is the immediate as well as the ultimate cause of *all* phenomena. This position, though implicit in Boyle, Clarke, and several other writers, was not easily upheld by a mechanical philosophy which so emphasised the rule of laws. Perhaps the best expression of it is by John Wesley,

who characteristically placed much emphasis on scripture. Speaking of the earthquake at Lisbon he wrote:[34]

> If by affirming, 'all this is purely natural', you mean, it is not providential, or that God has nothing to do with it, this is not true, that is, supposing the bible to be true. For supposing this, you may discant ever so long on the natural causes of murrain, winds, thunder, lightning, and yet you are altogether wide of the mark, you prove nothing at all, unless you can prove, that God never works in or by natural causes. But this you cannot prove, nay none can doubt of his so working, who allows the Scripture to be of God. For this asserts in the clearest and strongest terms that *all things* (in nature) *serve him*: that (by or without a train of natural causes) he 'sendeth his rain on the earth', that he 'bringeth the winds out of his treasures,' and 'maketh a way for the lightning and the thunder:' in general, that 'fire and hail, snow and vapour, wind and storm, fulfil his word.' Therefore, allowing there are natural causes of all these, they are still under the direction of the Lord of nature. Nay, what is nature itself but the art of God? Or God's method of acting in the material world? True philosophy therefore ascribes all to God.

It is also noteworthy that a similar position was taken by Leibniz when, in correspondence with Newton, he criticised the great Englishman for his reduction of the role of God to that of a cosmic mechanic. In other ways, however, they differed profoundly.

As we shall see in the next chapter Newtonianism was by no means the only system of natural theology that explicated in a clear manner a relation between God and nature. Ecclesiastical interest in science, however it was to be interpreted, was far too persistent in England for much weight to be attached to the old traditional view that the churches 'had lapsed into a tolerant indifference' towards science.[35] In fact, both in England and abroad efforts were continually being made to move (in the words of Pope) 'from nature up to nature's God'. Particularly amongst those to whom scripture was making a new appeal, science was seen as something attractive and worthy, if not always to be followed with the intensity and single-mindedness of Newton and his followers. Countless examples exist in the literature. In Würtemberg a local parson, P. M. Hahn, constructed a mechanical model of the universe which was greatly admired by the Emperor Joseph II. We are told it 'contained a device for stopping it in the year 1836, when according to Bengel's calculations, the return of Christ and the beginning of the millennium was expected'.[36] In America the pastor Jonathan Edwards (eventually President of Princeton) was known for his scientific activities as well as for his Calvinist theology.[37] For many years the great evangelist George Whitefield maintained a close and cordial friendship with Benjamin Franklin, greatly admiring his work in electricity.[38] John Wesley himself could be found reading *Philosophical Transactions*, Priestley's *History of Electricity*, and Burnet's *Sacred Theory of the Earth*, while when visiting Teddington in 1753 he spent an evening with Stephen Hales, saw some experiments, and concluded 'how well do philosophy and religion agree, in a man of sound understanding!'[39]

4. Nature and Society

4.1 'HIS MARVELLOUS PROVIDENCE'

IN August 1690, shortly after the Glorious Revolution and the ascent to the throne of William and Mary, two sermons were preached before the Queen by John Moore (1646–1714), her chaplain and a Canon of Ely. Addressing her 'on the wisdom and goodness of Providence' he observed that 'when God had made the world, he did not leave it to shift for itself, without any farther regard of it. But his power does as truly appear in the preservation and government thereof, as it did in its creation'. Similarly in political life 'God governs the world, and ordered all the affairs thereof'.[1] The Queen should have been well satisfied on this point for five months later she was informed by the Bishop of Worcester that the Revolution bore so clearly 'the marks of God's hand' that 'there is enough, one would think, to convince even the atheist to the belief of a Providence'.[2] On another royal occasion fifty-six years later Bishop Butler celebrated the twentieth anniversary of the accession of George III with a sermon to the House of Lords at Westminster Abbey. Speaking of the desirability of 'a quiet and peaceable life' he suggested to their Lordships that:[3]

> In aid to this general appointment of Providence, civil government has been instituted over the world, both by the light of nature and by revelation, to instruct men in the duties of fidelity, justice, and regard to common good, and

> enforce the practice of these virtues, without which there could have been no peace or quiet amongst mankind; and to preserve, in different ways, a sense of religion as well as virtue, and of God's authority over us. . . . Civil government has been, in all ages, a standing publication of the law of nature.

Between these occasions the view that social institutions were providentially ordained had become a conspicuous feature of the English intellectual tradition. Though much older than the Glorious Revolution, it was the accession to the throne by William and Mary that embedded this belief in the English statute book. For the Bill of Rights asserted that 'it hath pleased Almighty God in his marvellous providence and merciful goodness to this nation to provide and preserve their said majesties' Royal persons most happily to reign over us'.[4]

All of this may of course be interpreted as political rhetoric, bolstered and amplified by a subservient clergy. It seems to have little or no relationship with nature and even less with science. Yet this time there was talk of a miraculous East wind which brought the Dutch fleet safely to Torbay and to a bloodless coup. It was reminiscent of the 'protestant wind' that wrecked the Spanish Armada off the nothern shores of Britain just a century before. Moreover, it was precisely at this time that Isaac Newton was cogitating over the role that providence could play within the mechanistic universe described in his *Principia* of 1687. The issue for him was not whether God's providence was active in the frame of nature (he quite clearly believed that it was) but rather the manner of its action – by direct causes or by indirect? through gravity or in many other ways? Possibly, he wondered, comets might be the means used by providence to renew the universe from time to time.[5]

Thus in the concept of providence there is similarity of reasoning in both science and politics, what we might call an isomorphism. In fact the resemblances between the two fields went well beyond that. The universe as seen by Newton and his followers, was, despite the occasional 'interventions' of Providence, a very stable and ordered system, a model, it seemed, of the society for which men yearned after all the wars and revolutions of the previous century. Indeed, one could go further still and observe that, just as religion had been manifestly the inspirer and director of much seventeenth-century science, so now it was to be designated as 'this cement of society'.[6] Or, to change the metaphor, it could be seen as the bulwark against those tremendous forces that would tend to undermine and overthrow the whole structure of society.

These parallels between nature and man's social institutions were often referred to in the early years of the Enlightenment. Thus an unknown author wrote for the *Spectator* in 1712:[7]

> Nature does nothing in vain: the Creator of the universe has appointed every thing to a certain use and purpose, and determined it to a settled course and sphere of action, from which if it in the least deviates, it becomes unfit to answer those ends for which it was designed. In like manner it is in the dispositions of society, the civil economy is formed in a chain, as well as the natural; and in either case the breach but of one link puts the whole in some disorder.

At the same time the continued advocacy of 'natural religion' reinforced the notion of a congruence between nature and society. Although this also was no novelty in the eighteenth century, it takes to its ultimate conclusion the ancient view of God's revelation through the book of scripture and the book of nature.[8] If in time natural religion was never quite able to replace revealed religion in the minds of men it certainly was used in the early eighteenth century to validate biblical revelation. Natural theology, though a part of natural religion, is by no means all of it, for the proponents of natural religion were convinced that the laws of ethics were on a par with the laws of nature. To flout an ethical law was comparable to placing one's finger in the fire and hoping no damage would result. The tendencies are most clearly seen in Bishop Butler's *Analogy of Religion*, of 1736, where virtue is seen as a reasonable response to the way things are in society, and vice as a manifestation of unreason.

Given merely an isomorphism between views of nature and society, it by no means follows that any organic connection exists between the two. That must be a matter for detailed historical enquiry, and cannot be assumed. In the rest of this chapter we shall explore various links that have been suggested, and attempt to evaluate the evidence for each.

4.2 NEWTONIANISM AS A POLITICAL WEAPON

It has recently been urged that the science of Isaac Newton and his colleagues was taken up by the Latitudinarian wing of the Anglican Church after the Glorious Revolution and it was then used to justify their cause by providing the basis for a new and powerful natural theology. By this means, they hoped to rout the unbelieving factions that were seen to threaten national stability after the Revolution. At this time about eighty per cent of Anglican clergy were High Church Tories. As such they were committed to the divine right of kings and therefore opposed to the deposition of James II. The remaining twenty per cent, sometimes called Latitudinarians (because they included a fairly broad spectrum of churchmanship), were able to embrace Newtonianism with fervour for two main reasons. First it enabled them to take over concepts of providential intervention and thus justify the abrogation of divine right. Secondly, it gave them a most powerful tool for theological apologetics (the argument from design), thus enabling them to shine in public esteem as the upholders of religious values and supporters of the institutional structures of the church. These were coming under attack from various revolutionary sects with atheist leanings as well as dissenters and Roman Catholics.

This connection has been most cogently argued by Margaret Jacob, especially in her book *The Newtonians and the English Revolution, 1689–1720*. She claims that in the hands of the Latitudinarians 'Newton's natural philosophy served as an underpinning for the social ideology developed by the church after the Revolution'.[9] This was chiefly accomplished by their many exercises in natural theology, and above all by the Boyle Lectures. Here, then, was a possible mechanism whereby in the early years of the eighteenth century, science and society could interact with important consequences for each. But this claim, which is repeated frequently throughout the book, needs substantiation. We may summarise the case in four propositions:

(a) Many personal links existed between Latitudinarian churchmen and practitioners of Newtonian science.
(b) In the writings of the Latitudinarians there are many specific references connecting scientific and social issues.
(c) The intention of many Latitudinarians was to use their science to underpin the social order.
(d) The effect of these activities was to strengthen the social position of the church and to popularise Newtonian science, thereby promoting its general acceptance.

We shall examine each of these propositions in turn.

(a) First, there is the question of personal links between Latitudinarian Churchmen and Newtonian scientists. In the years before the Revolution numerous young clergymen with Low Church tendencies were involved with Newtonians in networks of friendship and kinship. Thus John Tillotson (1630–1704), who became Archbishop of Canterbury after the Revolution, was a Fellow of the Royal Society, son-in-law of John Wilkins, and friend of Isaac Barrow (Newton's teacher) and Edmund Halley (the astronomer). Another Archbishop of Canterbury, Thomas Tenison (1630–94) and the diarist John Evelyn (1620–1706) were both Latitudinarians, friends of Robert Boyle and custodians of his Lectureship. Edward Stillingfleet (1635–99), Bishop of Worcester, was a patron of Richard Bentley, the first Boyle Lecturer. Above all nearly all of them were personal friends or at least acquaintances of Isaac Newton himself. Thomas Sprat, early historian of the Royal Society (and 'trimmer' if ever there was one), was an ardent advocate of both the current social order and the fact that 'science needs princes' for its support. He also was a close friend of many of the Latitudinarian clergymen.

All this was of course before 1700 but alliances between Newtonians and Latitudinarians persisted well after the date, and they certainly can be discerned after the Boyle Lectures finished in 1717. A difficulty in maintaining the thesis up to (say) 1720 (the terminal date of Dr Jacob's book) is that after the Revolution ecclesiastical politics became much more complicated. The term Latitudinarian (like the term Puritan before it) becomes very much less clear with the passage of time. The later Boyle Lecturers were far from homogeneous in terms of their ideologies. Clarke and Whiston suffered the extreme disgrace of being charged as Arians, Whiston even being dismissed from his post at Cambridge.[10] Nevertheless substantial evidence does exist for an alliance in the eighteenth century between what might be called the second generation Newtonians and churchmen who did not espouse the Toryism of the High Church Party. That much is clear.

(b) For a detailed understanding of this network of relationships we are indebted more to Margaret Jacob than anyone else. Perhaps her major achievement has been to establish beyond any doubt the morphological similarities in arguments about society and about nature in the writings of moderate churchmen, and to show many cases of their specific connection. She observes 'the social philosophy of the Latitudinarians could rest on their understanding of the natural order, on the discoveries of the new science, because they believed that the worlds natural and political were interrelated'.[11] This harmonises exactly with the sentiments of Isaac Barrow:[12]

> As in the world natural, the parts thereof are so fitted in varieties of size, of quality, of aptitude to motion, that all may stick together. . . . And all co-operate incessantly to the preservation of that common union and harmony which was there intended; so in the world political we observe various propensions and attitudes disposing men to collection and coherence and co-operation in society.

A manuscript note by Newton himself refers to the mystical language of scripture as being founded on 'the analogy between the world natural and the world politic'.[13]

Given such an analogy one could legitimately argue from one 'world' to the other, and in particular could justify the existing social order as being, like the natural order, something ordained from above. But according to Samuel Clarke:[14]

> What the Sun's forsaking that equal Course, which now by

> diffusing gentle warmth and light, cherishes and invigorates everything in a due proportion through the whole System . . . would be to the *natural World*; That very same thing, Injustice and Tyranny, Iniquity and all Wickedness, is to the *moral and rational* part of the Creation.

Interestingly, he regards the moral law as more enduring than the natural law: 'the only difference is this: That the one is an *obstinate and wilful* Corruption, and most *perverse* depravation of Creatures made after the Image of God . . . whereas the other would be only a revolution or change, of the *arbitrary and temporary* frame of Nature'.

Nor was the analogy merely a question of the operation of predictable laws. We may be in for some surprises. As Clarke observed:[15]

> Tis exceedingly reasonable to believe, that *as* the Great Discoveries which by the Diligence and Sagacity of later Ages have been made in Astronomy and *Natural* Philosophy, have opened surprizing Scenes of the Power and Wisdom of the Creator, beyond what Men could possibly have conceived or imagined in Former Times: *so* at the unfolding of the whole Scheme of Providence in the Conclusion of this present State, Men will be surprized with the amazing Manifestations of Justice and Goodness, which will then appear to have run through the whole Series of God's Government of the Moral World.

That some of the Boyle Lecturers' vindications of social morality were practical in intent follows from specific assaults on the doctrines of Hobbes, whose subversive atheism was seen as a significant threat to the stability of the social fabric. Clarke expends pages of vituperation on 'Hobbes' state of nature and equality', and Bentley argues that sermons were to be chiefly directed against evil manners rather than evil books.[16] He too saw Hobbes as the chief target to be attacked, because his atheism corroded away 'the cement of society'.

There is little to be gained by further repetition of passages like these, but they are sufficient to justify a clear and conclusive affirmation that in the writings of the Boyle

Lecturers and others nature and society are treated as parallel cases of God's overruling providence. That by itself is not sufficient, however. The association of political and scientific ideas in individuals, groups or in published books does not necessarily mean that science was used to legitimate social structures. It could equally well be the other way round, and in some cases it probably was. When science in the early Royal Society was being ridiculed or even ignored it needed someone of the intellectual stature and verbal dexterity of Thomas Sprat to justify it as a serious enterprise. One could go further. A conjunction of science and politics need not be held to be a vindication of either, particularly in the context of natural theology. There is always at least the possibility that each was used for the support of religion *per se*.

(c) So we need further evidence, and Dr Jacob has attempted to supply this. She writes:[17]

> The conclusion becomes inescapable: if we are to discover why these churchmen adopted a world view based upon what they made of Newton, we must look to them for an answer. Their Boyle lectures reveal that they consciously rejected alternative natural philosophies in favor of the Newtonian system. What is more, they pitted this 'new and invincible' system against the mechanical philosophy of Hobbes, the fatalism of Epicurus, the remnants of Aristotelianism, and the alternatives posed by contemporary freethinkers.

That is indeed incontestable. But her paragraph continues:[18]

> The Newtonians did so not simply because they disagreed with these explanations of how the universe worked, but because they also and primarily saw these philosophies as profound threats to the social, political, and religious order – the basis they imagined for the security of church and state. The Newtonian model provided for these churchmen a foundation upon which that order might rest secure, and for that reason Newton's philosophy proved irresistible.

It is at this point that the argument acquires a new note which, for all its confident ring, leaves us feeling slightly uneasy. The author moves from demonstrable fact to speculation, but in such a graceful manner that the change of key is not obvious. We are told that all this happened because the churchmen imagined (or knew) that the existing social, political, and religious order was under threat, and that this was one way to encounter the foe. Statements of this kind occur again and again in the book, but repetition does not constitute an argument, and little is quoted that unambiguously shows that the Boyle Lecturers were *primarily* interested in supporting a social or political ideology grounded upon Newton's science. The mere conjunction of social and scientific ideas is not, as we have seen, sufficient evidence. Nor is the explicit making of political points, unless this is manifestly the principal thrust of the whole strategy. At times this seems to be perilously near a cyclic argument. In defence of the view that Newtonianism cannot be 'transcendent of ideology',[19] Dr Jacob opposes the assertion that Newtonians simply restated emphases of thirteenth-century Christian theology on the grounds that 'such an approach rests on the methodological assumption that the ideas of the Newtonians could and did exist independent of, or isolated from, a prevailing social, political, and economic environment'.[20] This is suspiciously like saying that Newtonianism is socially conditioned because, in the last analysis, it *must be* socially conditioned!

The political aims of the Boyle Lecturers have been defined more precisely as the justification of a Christianised capitalism, expressed in the pursuit of self-interest.[21] Elsewhere Jacob speaks of a domination by 'market values' of the Latitudinarian ideology. She believes that the chief achievement of these Newtonians was 'to synthesise the operations of a market society and the workings of nature in such a way as to render the market society natural'.[22] Undoubtedly 'market values' may be found in the Boyle sermons but it is possible to put an entirely different construction on their appearance. Granted that many in the London congregations were prosperous merchants, granted that Newtonians did not condemn self-interest as such but rather its disorderly abuse, granted even the frantic run of financial speculation that led to the South Sea Bubble of 1720, it is still perfectly possible to interpret the numerous allusions to a market economy as apposite illustrations in an address designed for quite different ends. Latitudinarians knew full well that they had to adapt their ministry to their audience. Isaac Barrow had bemoaned the inappropriateness of the Anglican liturgy for 'men of business and dispatch'. It suited 'Cloisters (whence so much of it came) rather than congregations of tradesmen and merchants'.[23] The Boyle Lecturers were skilled preachers; they knew their audience and knew the illustrations that would most rivet their attention. Naturally they made the most of their opportunities. Jacob's remarks about deists and atheists could equally well be applied to them: 'they sought support for their arguments wherever they could find it – in science, philosophy, political theory . . .'.[24] She quotes from a passage from Derham as his 'assessment of the business world',[25] but in fact it occurs in a passage devoted to the keeping of Sunday and appears as a most natural illustration and reinforcement of his argument:

> Thus the wise Governour of the World, hath taken care for the Despatch of Business. But then as too long Engagement about worldly Matters would take off Mens Minds from God and Divine Matters, so by this Reservation of every Seventh Day that great Inconvenience is prevented also.

When a thesis of this kind is applied to the period with which we are primarily concerned (after 1700) it is susceptible to further objections. In an essay review of Dr Jacob's book, Professor G. Holmes has observed that the 'monied' threat to the established hierarchy was largely confined to the London area and that it receded after 1713; that even in the 1730s society was largely dominated by the established landed families rather than city financiers, and that 'the élite group of enlightened divines . . . so far from being *most* susceptible to the bogy of the 'monied' interest, were almost certainly the *least* likely element among the clergy to be alarmed by it. After all, it was the Latitudinarian and the Low Church clerics who, for the most part, shared the Whig politics of the monied men'.[26]

There simply does not seem to be unequivocal evidence for asserting that the Boyle Lecturers were primarily concerned to advance a social or political ideology. It would therefore seem better to see, with Schofield, the adoption of Newtonian natural philosophy by Christian apologists as a response to the challenge of atheism.[27]

To object that this ignores 'the ideological uses these churchmen then extracted from Newtonianism'[28] is to beg the whole question, which is whether there were any ideological uses, other than the general wish to defend a faith apparently under siege. Jacob finds it incomprehensible that the conflict was 'centred essentially on intellectual issues'[29] or that it was 'only simple Christian piety that was at stake'[30] but to deny these possibilities is just another facet of the assumption that allegedly theological arguments must of necessity be socio-political manifestos in disguise.

It appears, then, that Newtonianism was most certainly adopted as an integral part of the Latitudinarian creed and that its resonance with political claims made it an attractive and compelling part. But it does not therefore follow that the socio-political content of some of the Boyle Lectures was the rationale of their existence. Unambiguous evidence is needed, and has not yet been supplied.

(d) If, as we have argued, there are no substantial grounds for asserting that the later Boyle Lecturers at least were intent on underpinning the social fabric with Newtonian science, it still does not follow that the thesis falls to the ground. It could be the case that, whatever the intentions might have been, the effect of the Boyle Lectures was to achieve precisely the results described, and, almost accidentally perhaps, to stabilise Augustan society. For this to be true we might expect to find evidence that the Boyle Lectures had immediate, considerable and lasting effects and that these could not be attributed to any other causes. To be sure, the Lectures attracted large congregations and some were repeated time and again. Yet, as Holmes pointed out, they cut little ice with the majority of clergy, were outlived by the deism they hoped to quell, and were unnoticed in the voluminous diaries of William Nicholson, Bishop of Carlisle.[31]

Further, if Newtonian propaganda had succeeded in its alleged socio-political objectives it follows that Newtonianism itself must have acquired new stature in the eyes of many people. It is widely held that Newtonian ideas did spread rapidly in England, but there were plenty of other ways in which this could have happened. One need only refer to the treatises of Maclaurin, Desaguliers and others; to the popular lectures in London and elsewhere by many itinerant speakers, including Desaguliers in 1713; to the countless conversations in coffee houses and other haunts of polite society; even to lectures at Oxford and Cambridge. It is noteworthy that as Gillispie points out, authors in natural philosophy rarely attempted to 'draw any social or political inferences from their subject'[32] – which is surprising if Newtonianism and socio-political ideologies had become so intimately connected. One possible exception to this generalisation is Desaguliers himself whose *The Newtonian system of the world, the best model of government*, however, was just 'a conventionally elegant *jeu d'esprit*' to celebrate the coronation of George II.[33]

But Newtonianism did not only spread in England. In Presbyterian Scotland and in Calvinist Holland it flourished mightily in the early decades of the eighteenth century, and it is hardly possible to credit this achievement to the exertions of the Boyle Lecturers. As Roy Porter says in a review of Jacob's book, before the nineteenth century science was often 'a rather minor weapon in the ideological armoury . . . a weapon particularly likely to backfire'.[34] As we have seen, and shall see again, natural theology was able to hold its own quite effectively without a close connection with any obvious political progress. What Newtonian science was able to do in the period under review, however, was to give a substantial bolster to Christian apologetics, and to turn it in a particular direction that had the profoundest effects in the later history of science. In fact the effect of post-Newtonian natural theology was to produce an impressive alliance not between science and society but between science and scripture. As Manuel says 'the Bible and the new science were being locked in deadly embrace',[35] one result of which was a plethora of books in English deducing the future and past history of the earth from Biblical data exclusively. While we may not agree with him that 'during the first decades of the eighteenth century the glorias of the Boyle Lectures reached unprecedented levels of banality',[36] it is certain that they plumbed unprecedented depths of detailed natural history. But their immense popularity, as evinced by the flood of similar publications through the eighteenth century, can leave us in no doubt that in England, at least, Newtonianism owed a good deal of its popularity to the Boyle Lectures, their imitators and successors.

Finally, it is worth briefly noting that on the continent a similar form of natural theology was able to develop, though the European tendency to separate science and scripture (as in the earlier cases of Galileo and Kepler) was largely maintained. According to Henry Guerlac, Newtonianism was popularised in Europe by earlier and quite different means. The publication in 1672 of Newton's first letter on light and colour, his invention of the reflecting telescope and the propaganda war assiduously waged on his behalf by Oldenburg seemed to have been the chief modes of popularising Newton at the end of the seventeeth century. Later, of course, he was 'reintroduced' into France by Voltaire whose sharp polarisation between Newton and Descartes is, in Guerlac's view, overdrawn. He instances the followers of Malebranche who occupied an intermediate position and helped to break down the internal defences of Cartesianism.[37]

4.3 SOME ALTERNATIVES TO NEWTONIANISM

While the Latitudinarians were using Newtonianism in their defence of orthodox belief, their opponents continued to flourish in the early years of the eighteenth century. This is no place to attempt an analysis of the multitudinous sects, ranging from slight deviations from theological orthodoxy to the fringe and beyond. It is sufficient to note that the flood of deist literature may well have been seen by many churchmen as posing a threat to political stability, though that does not necessarily mean that their natural theology was conceived with that primarily in mind. The claim by Toland merely to belong to a secret society was enough to goad the nervous into greater anxiety, even though it appears to have been mainly concerned with ritual celebrations of Pantheistic philosophy. Toland's rejection of inactive matter, and thus the foundation of Newton's natural philosophy, was nearer to Pantheism than Atheism, but was unhesitatingly regarded as the latter by Newton and his disciples. Again there is an isomorphism between Toland's universe, free of divine interventions, and a political world that runs itself. Free thinking certainly seemed to be coupled with republican politics.[38]

While deism was still flourishing it was accompanied by various movements that tended towards Arianism, i.e. a variegated cluster of beliefs that denied in one form or another the full doctrine of the Trinity as expounded by the Thirty-Nine Articles and Athanasian Creed. Newton himself was not above suspicion in this matter, and his notebooks have since disclosed how deep his forebodings were about the full pre-existent deity of Jesus Christ. His less tactful and less prestigious followers who shared his doubts fared less well, as was the case with Whiston. Unitarianism, as a coherent set of beliefs that stem logically from these doubts, emerged gradually in England during the eighteenth century, chiefly amongst dissenters and especially the Presbyterians. Its most notable scientific representative was Joseph Priestley, who by 1755 had begun the short pilgrimage from Arianism to Unitarianism. In conformity with Priestley's denial of the deity of Christ, as well as of immortality and free will, came the abandonment of the matter–spirit distinction, and the denial of solid atoms.[39] Moreover Priestley was almost the only English scientist of any distinction to adopt a radical political stance. This factor, coupled with his theological heterodoxy and his popularity in France, goes far to explain the burning of his house in Birmingham by a mob in 1791. Once again science, society and religion were seen to be locked together in the life and experience of one man.

Against deism, atheism and Arianism the English established church attempted to stand firm. And in doing so, as we have seen, the party ascendant early in the eighteenth century – the Latitudinarians – used the science of Isaac Newton and his followers. But what of the other wing of Anglicanism, the High Church? They were outnumbered amongst the Bishops, but they still had the majority amongst the ordinary clergy, the so-called Lower House of Convocation. It is only in the last few years that a clear picture has emerged of an alternative strategy adopted by these churchmen, a strategy which also used arguments from nature but which rejected Newtonianism. This new ideology has been termed Hutchinsonianism[40] after its founder John Hutchinson (1674–1737), Steward to the Duke of Somerset. His colleague, the Duke's physician, was John Woodward, who was credited with introducing him to natural philosophy. Impressed by the fossils dug up during his survey of the ducal estates, and inspired by Woodward's interest in the Mosaic account of creation, Hutchinson turned to Biblical studies and to the original Hebrew. He concluded that fire, light and air are material agents corresponding to the three Persons of the Trinity, and this became a corner-stone of his natural philosophy. Had he but known it, his Trinitarian emphasis alone would have distanced him a very long way from the system of Isaac Newton. In fact the two systems differed in several other more obvious ways. Hutchinson took issue with Newton's insistence on action-at-a-distance without any mechanical causes. He believed that an ethereal fluid transmits forces, thus denying Newton's 'void', or vacuum, in space. The supposition that Newton's immaterial agent might be God struck Hutchinson and his followers as tantamount to nature worship, and almost to Pantheism. They stressed the Biblical notion of a God who was transcendent, i.e. existing independently, of his universe. Occasional interventions in the Newtonian manner were not required.

Hutchinson's natural philosophy became known in the 1730s, and gained many influential supporters in England and Scotland. In Oxford one may include Walter Hodges, Provost of Oriel College, George Horne, President of Magdalen and Vice Chancellor, and William Jones (biographer of Horne and author of numerous works on Hutchinsonian physico-theology, including his essay on the *First Principles of Natural Philosophy* of 1762, *Physiological Dispositions* of 1781 and other purely theological works). It survived until well into the nineteenth century. What is important now is the fact that Hutchinsonianism was adopted by the High Church party in response to the challenge from the Latitudinarians, and even when that party disappeared as a viable force individual members of it continued to promulgate Hutchinsonian doctrines. They had much to commend them. First, Hutchinsonianism avoided the heretical overtones implied by much Newtonianism, despite the early stand against deism taken by many followers of Newton. Thus Clarke and Whiston had been discredited for their Arian leadings, and the notorious Toland had once been a student of Newtonianism in Scotland. Later on in the century one only had to look to the *philosophes* in France to see where Newtonian natural philosophy could lead a nation. Secondly, this view of the universe that did not need to receive continual divine interventions meant a world view more acceptable to High Churchmen whose belief in divinely ordained kingship could not be so easily blown away by the 'protestant wind' of an overriding and overruling providence. Thirdly, the Pantheist overtones detected in some of Newton's references to God in nature needed a corrective, and this was ready to hand in the emphasis by the Hutchinsonians on the transcendence of God. Fourthly, the downgrading of revelation in late Newtonian natural theology was repugnant to the High Anglicans; as one modern Anglican has put it,[41] though with a measure of overstatement:

> Neither in Butler nor in his contemporaries is there any awareness at all that the heart of Biblical religion is the proclamation of the saving acts of God in history, or that Christianity began as the proclamation of an historical event and not as the republication of the religion of nature.

Many of the ordinary clergy in the early eighteenth century (who were in the main inclined to the High Church) would view the Newtonian exercises in natural theology in very much the same way. Hutchinson, by avoiding such theological regression, would be preferable, they thought. Fifthly, there was a specifically Hutchinsonian emphasis on hereditary kingship and a stratified society which, though originating in the linguistic studies of the Old Testament, was naturally attractive to a Tory mind. Sixthly, the Hutchinsonians tended to belittle the cult of reason and to stress the need for experience and (above all) revelation. In these and in other respects their *credo* was completely consistent with the dominant Augustinian theology of the High Church, which attached particular importance to the majesty and sovereignty of God, especially as manifest in the Incarnation. The morphological similarity between Augustinianism and empiricism has been noted by Donald Greene,[42] and Wilde has shown how his analysis 'fits Hutchinsonianism perfectly'.[43]

All these factors make the association between Toryism and Hutchinsonianism readily understandable. This new natural philosophy received a boost from the discoveries in electricity around 1750, for these seemed particularly susceptible to an ethereal interpretation.[44] The importance of these ideas for Daltonian atomism has been emphasised by Thackray.[45]

Hutchinsonianism was to be found among the Episcopal clergy in Scotland in the eighteenth century, a circumstance seemingly attributable to the influence of John Skinner (1721–1807), Dean of Aberdeen, and his identically named son (1744–1816), Bishop of Aberdeen. In France Hutchinson seems to have cut little ice on account, we are told, of the 'air of mysticism . . . and the presumptuous tone of the author', so much so that 'it is often enough to arrest the advancement of a man of merit by presenting him as a Hutchinsonian'.[46] In England, however, his popularity was by no means confined to High Anglicans. The evangelical revival, particularly in the case of the Wesleys, had little in common with the rational cult of nature so commonly associated with Newtonianism. Did not Charles Wesley sing 'Long my imprisoned spirit lay/Fast bound in sin and *nature's night*'?[47]

In fact John Wesley urged his students to study Hutchinsonian doctrines.[48] When, on the 9 October 1763, he read William Jones' *First Principles of Natural Philosophy* he observed 'he seems to have totally overthrown the Newtonian Principles. But whether he can establish the Hutchinsonian, is another question'.[49] His last remark shows the breadth of his reading, since in this book Jones is careful to avoid mentioning Hutchinsonianism. Writing in a tract 'Serious thoughts occasioned by the earthquake at Lisbon', Wesley was careful to assert that 'natural causes' – not just occasional interventions – 'are still under the direction of the Lord of nature',[50] a position that Newton may have allowed but not some of his later followers. At the very least Hutchinsonian physico-theology gave to Wesley and his friends a confidence that their faith was not incompatible with scientific enquiry, and to that extent may have played some small part in a movement that was to have a profound effect on English and American society.

Finally, to emphasise the dangers of over-simplification and too rigid a categorisation of 'parties', we briefly examine the case of one man who, in mid-eighteenth century England, sought to relate the findings of science and social issues of the day, and did so by means of natural theology. He was James Hervey (1713–58) Anglican Rector of Weston Favell, near Northampton, whose *Meditations*[51] and *Contemplations*[52] went through many editions in a very short time both in England and abroad. These works had a stronger emphasis on natural theology than some of his later books, though these were equally popular.

Hervey studied at Oxford from 1733 to 1736, just a few years before the Hutchinsonian influence became strong in that university. Like Hutchinson himself he was to spend much time on the detailed study of the Hebrew language. He was introduced to the standard works on natural theology of Ray, Derham, Keil and others, and acquired a strong admiration for Newton (according to one account, he saw, 'with intuitive rapidity of perception, that gravitation is only another name for the action of God our Saviour, upon all matter in the universe').[53] Here he made the acquaintance of the Wesleys and Whitefield, fellow students at Oxford. Following a correspondence with Whitefield be became a convinced evangelical (1741) and shortly afterwards set to work on his natural theology.[54]

Partly through ill-health Hervey's subsequent life was outwardly uneventful and largely confined to writing and his work as a minister. In his natural theology there are many purple passages of a conventional kind of which one example is quite enough:

> [God] rounded in his palm those dreadfully large globes which are pendulous in the vault of heaven. He kindles those astonishingly bright fires which fill the firmament with a flood of glory. By him they are suspended in fluid aether,

and cannot be shaken: by him they dispense a perpetual tide of beams, and are never exhausted. He formed with inexpressable nicety that delicately fine collection of tubes, that unknown multiplicity of subtle springs which organise and actuate the frame of the minutest insect. He bids the crimson current roll, the vital movements play, and associates a world of wonder, even in an animated point. In all these there is a signal exhibition of creating power; to all these are extended the special regard of preserving goodness. From hence, let me learn to rely on the providence, and to revere the presence, of the Supreme Majesty.[55]

Newton and Galileo are praised,[56] the Northern lights,[57] the phases of Venus,[58] the wonders of plant life are all displayed,[59] and, true to the spirit of natural religion, he asks,[60] whether the 'necessity' of natural laws should not be paralleled by the 'necessity' of moral obedience. Clearly there is much of Newtonianism in all this, though the reference to 'fluid aether' in the quotation is perhaps more suggestive of Hutchinson.

Various passages reinforce the notion that Hervey is trying to use nature to justify the social order. With the recent memory of the 1745 disturbances he writes of 'liberty that dearest of names and property that best of charters'.[61] One could hardly be more political than that. Reflecting on the adaptation of flowers to the environment he exclaims 'what a striking argument is here for resignation, unfeigned resignation, to all the disposals of Providence!'[62]

But Hervey is not just being a tool of the Establishment. He trenchantly condemns the godlessness of 'many of the most conspicuous families' as displayed in 'our genteel interviews'.[63] And in 'A winter piece' he pours the most scornful invective on to the rich:[64]

While the generous juices of Oporto sparkle in your glasses, or the streams beautifully tinged and deliciously flavoured with the Chinese leaf, smoke in the elegant porcelain, O remember that many of your fellow-creatures, amidst all the rigour of these inclement skies, are emaciated with sickness, benumbed with age, and pining with hunger.

Passages like this neutralise the impression of a political tract, or at least one directed from the Establishment downwards.

But they tell us little of Hervey's real purpose. This, however, may be discerned on almost every page, as primarily to arouse men's minds and consciences to the call of the gospel, to repentance and faith. After that, appropriate social action should follow. As he says:[65]

> I think we should always view the visible system with an evangelical telescope (if I may be allowed the expression) and with an evangelical microscope, regarding Christ Jesus as the great projector and architect, who planned and executed the amazing scheme. Whatever is magnificent or valuable, tremendous or amiable, should ever be ascribed to the Redeemer. This is the Christian's natural philosophy.

He wonders at the vulgarity of minds that cannot discern the implications of modern astronomy,[66] and, like Wesley, sees the hand of God in 'natural' disasters.[67] Here is natural theology in the spirit (if not the actual practice) of the Boyle Lectures, but with a strong evangelical emphasis. It goes one step further as Hervey regards the wonders of redemption as even more marvellous than the wonders of creation. Significantly, he thinks that this might be deemed an innovation without precedent, and hastens to assure us that it is not quite the case.[68] In fact this is one of the earliest instances of natural theology turned consciously and deliberately in a new direction. It was no longer concerned to alter (or conserve) the structures of society, for it now had a much more radical programme: nothing less than a transformation of the individual human heart in response to the grace of God. And that, he believed, would really go to the root of social evils.

5. The Social Organisation of Enlightenment Science

5.1 ROYAL SCIENCE

THERE is no 'inevitability' about scientific progress. The very variable rates at which science progressed during the Enlightenment – and at other times – should be enough to convince us of this. Some have argued that few scientific theories are entirely value-free and that the practice of science depends greatly upon the cultural context in which it flourishes or wilts. At the very least social changes will affect the rate of scientific advance (leaving aside the question of its direction). The purpose of this chapter is to explore some of the ways in which Enlightenment science was affected by its social framework. In the broadest sense of the term we are talking about the *institutionalisation* of science. Except for the hypothetical scientist working alone on his desert island, screened from all influences from the outside world, everyone who practises science does so in some kind of institutional framework, even though the institutions may not be primarily designed for science itself (as, for example, when the scientist works for a firm or pursues a strong amateur interest within a loosely structured local community). But the institutionalisation of science as conventionally understood refers to the formally constituted bodies which are dedicated primarily or exclusively to the practice of science. Such societies were not a product of the Enlightenment but rather of the previous century. None of them had – or has – greater prestige than the two great national societies of England and France, the Royal Society of London and the Académie Royale des Sciences. We start our account with these, not because of any primacy attributable to their 'Royal' status but because they represent the ultimate in institutional science during the Enlightenment.

Both societies were creations of social changes in Europe at the end of the seventeenth century. The origins of the Royal Society have been frequently discussed, not always with agreement.[1] However, no one doubts that they lay in the ferment of intellectual and scientific excitement in mid-seventeenth-century England, or that this was strongly expressed by the group of young men who associated with John Wilkins at Oxford. It is generally, though not unanimously, accepted that a London group meeting a few years earlier at Gresham College and elsewhere, constituted the original nucleus from which the Royal Society later grew. Their Baconianism was of little interest at Court, however, and the distinction of Royal patronage was unthinkable until after the Restoration. With the ascension of Charles II, the embryo society was anxious to avoid all suspicion that their practices might be tinged with 'enthusiasm', or worse. Amongst their sympathisers was Sir Robert Moray, a Privy Councillor known for his loyalty both to the new King and to his father. Within a few months of Charles's arrival in London in 1660 Moray had spoken favourably to the King about the proposal to form a Society, and conveyed to the delighted philosophers an assurance of Royal favour. Its first Charter was granted on the 15 July 1662. From then on one of its immediate objectives was to reinforce its 'safe' image, hence the so-called *History* of Sprat which deliberately played down the puritan sympathies of many of its founders; and hence also the flooding of the Society by members of impeccable ancestry and reputation.

Baconianism was also the inspiration behind the Académie des Sciences, founded in admiration of its English counterpart, anticipated by ambitious plans for a new Compagnie des Arts, and receiving a more realistic title at its inception on 22 December 1666. After a somewhat uncertain start it received new and more precise regulations in 1699, ratified by the French *parlement* a few years later.[2]

Characteristic of the aspirations of late seventeenth-century intellectuals were two features embodied, more or less, in all the scientific academies of their day. These features, to which Hahn has drawn special attention,[3] were (a) a preoccupation with open communication, and (b) a sense of cultural superiority. In the former the emphasis was on vernacular writing, frequent meetings, extensive correspondence and (by implication) a willingness to accept changes of opinion or organisation. In the latter the men of science became, or sought to become, a community with real power within a nation. This need not have been political power as ordinarily understood. It could have been – and sometimes was – what a later English chemist was to call:

> that great sublime unquenched love of fame,
> that love of fame that never, never dies.[4]

Or, as a member of the Berlin Academy, J.-H.-S. Formey put it, 'it is up to the Academy to establish sound knowledge, rich in precious fruit, which chases out demi-savoi', (the half-knowledge professed by ordinary people) 'this legion of half wits, not even worthy of being called dung piles . . . who dishonour the Republic of Letters'.[5] This sense of cultural élitism arose simultaneously with the new scientific academies. It coincides also with a growing sense of national identity and new modes of government in Europe. Science was clearly taking on some of the aspects of organised religion, anticipating Huxley's much later designation of the 'church scientific'.

As Hahn has put it, the academy (whether scientific or otherwise) was 'a dominant cultural institution of the late seventeenth century',[6] its role and function being allied to the centralising or bureaucratising tendencies of the government. Where, as in Italy, these were far from complete the 'republic of letters' might be seen as an instrument for national unification; where (as in France) absolutism was almost complete, knowledge was then seen as the ornament and tool of authority.

These considerations are well illustrated in the membership of the two societies. Originally the Royal Society was limited to 55 members but, with an eye on the political advantages of noble connections, it allowed that 'any person of the degree of Baron or above may be admitted as supernumerary, if they shall desire it'.[7] By the end of the seventeenth century, however, the total number of ordinary Fellows had reached 125. This was in striking contrast to the situation in Paris, although there were some similarities. Nobles and persons high in the ecclesiastical establishment were admitted as *honoraires*; they rarely attended scientific meetings, but would be frequently seen on social occasions. There was also a class of non-resident academicians who as *correspondants* or *associés* were for various reasons unable to attend meetings regularly. They included eight foreign members (far fewer than the Royal Society). But the biggest difference lay in the ordinary members of the Académie, those who had attended the bi-weekly scientific meetings regularly and formed the core of the establishment. Throughout the eighteenth century their numbers were kept at a virtually constant level, just over 50. Amongst many other things this means that membership statistics of the Académie cannot be used as an index of national scientific prosperity, whereas with most other societies some general conclusions usually can be drawn from such data. This restriction on membership indicates the determination of the Académie to maintain the highest scientific standards and to be itself the spearhead of scientific advance in the French nation. Indeed its membership showed a professionalism almost totally absent from the Royal Society. Unlike most counterparts in Europe the active membership of the Paris Académie was confined to specialists in one or the other of the sciences. Academic excellence and approval of the crown (not always coincident) were the two inviolable conditions for membership.

All this may be seen as an institutionalised expression of the respect shown by the Old Régime in France for science, and the determination of government to make its academy part of the royal administrative apparatus. No such determination existed in England where, if science was 'used' at all, it was only as an ideological justification of the existing social order, but even there only in a very minor way, and fairly early in the eighteenth century. Once their enterprise had been accepted by the King, Fellows of the Royal Society were only too anxious to be involved with the government as little as possible.

The 'professionalism' of the Académie, as contrasted with the amateur tradition in England, could find no better illustration than in the question of finance. An Academician was granted a pension by the crown as soon as he joined, but an FRS was required to pay his own way by a subscription (£2 12s.0d in 1660 raised to £4 in 1831) or by a system of composition fees. The fact that the Royal Society was in frequent difficulties over non-payment of dues (especially around 1740[8]), reflects its predominantly non-scientific membership, such members presumably finding social advantages not worth the money. On the other hand the deliberate waiving of fees in the case of a distinguished member may reflect a growing desire to attain a more professional image.[9]

Further reflecting the hierarchical structure of pre-revolutionary France, the ordinary membership of the Académie was stratified into 'classes' on the basis of seniority and social status. Eminent talent might secure admission in the first place; what happened after that to the new Academician would depend considerably on his own social position.

For all the period with which we are now concerned (roughly 1700–90) the Royal Society was predominantly an amateur body with the proportion of 'scientific' Fellows hovering around the 30 per cent mark. To take the figures for 1740: of the 100 scientific Fellows, 63 were medical men, 12 mathematicians, 8 botanists, 7 astronomers and 10 others. The figures available for 1770 are not significantly different.[10] Clearly Royal patronage of science meant two very different things in England and in France.

Membership of the French Académie produced other privileges in addition to the Royal pension. There were, at different times in the eighteenth century, tax advantages, probable exemption from military service, the honour of being presented to the King on admission to the Académie, the opportunity to wield real political power by membership of government technocratic committees, and above all the privilege of being amongst the very aristocracy of French intellectual life, the élite of the country's science. And of course one's scientific work could be discussed in the academic meetings and published in its journal.

By contrast the benefits conferred by Fellowship of the Royal Society were much less tangible. The honour of being associated, however indirectly, with Isaac Newton grew less as time wore on. The lapsing subscriptions told their own story. For the scientifically inclined there was the facility of attending meetings and publishing papers in *Philosophical Transactions*, but few other 'perks' were available even for the most senior of Fellows. Perhaps it was overseas that Fellowship seems to have been most prized. The numbers of foreign Fellows steadily increased from 1700 towards mid-century, and (more important) so did their proportion. In 1766 the Council decided that the proportion of foreign members was too high and took measures to reduce it; in future only 2 should be elected each year and the total number should eventually be reduced to 80 (ten years later that figure was amended to 100).[11] When Martinus van Marum sought election to the Royal Society at the close of the century he found it the hardest of all European academies to enter.[12] Perhaps the restriction on foreign membership may be seen as indicating a national pride that needed to be asserted in the face of French wars and (later) American independence.

Finally, we may enquire what was actually achieved by the two societies (as opposed to individuals in them). In the Paris Académie privilege also spelt responsibility, though at times the one merged into the other. For example the Académie was granted printing rights and permission to place its coveted imprimatur at the head of books 'approuvé par l'Académie Royale des Sciences'. In addition it regularly published its *Mémoires* for Academicians, and its *Histoire* for paraphrases of work done by others. Even the famous *Journal des Scavants* became, in 1702, its semi-official organ, with an editorial board exclusively composed of Academicians. In their exercise of such privilege (and responsibility) the Academicians necessarily had to apply censorship, hitherto a right of only the university and church, but now brought under the Crown (doubtless to its own great satisfaction). It also became the effective adjudicator of patent rights. In 1699 all technical projects had to go through the Académie which, through its committees, advised on the grounds of 'privilege', a temporary monopoly roughly equivalent to an English patent. In so doing it insisted on an adequate scientific basis for inventions, required detailed specifications, drawings and even models. Such requirements, however reasonable, were hardly conducive to technical enterprise amongst artisans and workmen much more accustomed to rule of thumb procedures. They served in the short run to increase still further the élitist reputation of the Académie, and, in the long run, to identify the whole hierarchical machinery of the old régime to be overthrown at the revolution. Moreover the exaltation of science at the expense of 'mere' technology seriously hampered the development of the French craft industry. As H. T. Parker observed of the French artisan, 'partly because he was devoted to the perfection of a handicraft process, he was unconsciously falling behind his English competitors'.[13]

The Académie was also required to advise the government directly, either through committees or through designated individuals. Chemical manufacture, civil and mechanical engineering, coinage, metrication and much else were subjects of official consultations. Its influence can be detected in almost every aspect of French technology and science – from the approval of shoe-cleaning materials to the appointment of a professor at Rheims and ratification of his proposed curriculum.[14]

By contrast the Royal Society was involved only sporadically with government until well into the nineteenth century. Usually it had to take the initiative and solicit government support, as when in 1784 it sought funds for a co-operative venture with the Paris Académie on the determination of longitudes. Individuals, however, might be consulted as when, in 1714, the President of the Royal Society was appointed one of the commissioners for the new Board of Longitude, or in 1717 when Newton, as Master of the Mint, defined the parity figure for gold.[15]

There were marked differences in the output of scientific work of members. While the Paris Académie continued through the eighteenth century to produce successful and important papers, in the Royal Society it was far otherwise. After an initial spurt associated above all with Newton and Newtonianism, the Royal Society declined into a gentleman's club, with some perfunctory scientific interests. Lacking the monopolistic rights of the French Académie, it also lacked that body's 'spiralling prosperity'.[16] As early as 1710 its instrument collection was reported to be in a ruinous state; from 1741 to 1748 (between the presidencies of Sloane and Banks) it slumbered under the reign of several undistinguished presidents, one of whom, Martin Folkes, regularly went to sleep in the chair. Financial difficulties around 1740 were just one index of the misfortunes that had befallen the Society.

In conclusion we should note that France and England were not alone in nourishing prestigious scientific academies, nor were they even the first in the field. That distinction belonged to Italy with a few short-lived academies before the famous Academia del Cimento of Florence (1657).[17] By the middle of the eighteenth century scientific societies were found in most major European countries. Academies were founded in Berlin, Leyden, Uppsala, and other important cities. Scotland's Royal Society of Edinburgh did not appear until 1783 (of which more later) and the Royal Irish Academy was founded in Dublin in 1785. All of these were associated with important scientific developments, and each became, in varying degrees, a symbol of national pride. But none achieved the eminence which, despite their failures, was internationally accorded to the Royal Society of London and the Académie Royale des Sciences of Paris.

5.2 PRIVATE PATRONAGE OF SCIENCE

The cultivation of science as a gentlemanly occupation was no new phenomenon in 1700. In England especially it dates back to the young manhood of Robert Boyle and even beyond. With the coming of a new political stability after 1689, and the growing prosperity of many of the landed gentry, the eighteenth century became increasingly an age of leisure and pleasure, both for them and for the many less affluent citizens who nevertheless shared in the general prosperity of the age. The leisured Englishman was now able to supplement the more conventional social pleasures of theatre, concert, dance and so on with a variety of activities that could properly be called scientific.[18]

These activities took many forms. For those who could afford them the excitement of scientific experiments on a large scale was a source of limitless pleasure. Apart from the Honourable Henry Cavendish at the end of the century, few aristocrats were able to devote a great deal of their time to such work, but had instead to bask in the reflected glory of the science performed by those in their employment. Patronage of the arts was of course no new thing, but now science itself was becoming esteemed by several of the more enlightened members of English aristocracy as worthy of their personal support and encouragement. In the seventeenth century the astronomers Edmund Halley and John Flamsteed had been encouraged by Sir Jonas Moore, Master of the Ordnance, who provided assistance with instruments and recommendation at Court,[19] and this kind of general encouragement continued to be shown from time to time. With the development of political patronage on a hitherto unprecedented scale after Walpole's appointment of the Duke of Newcastle as Secretary of State in 1724, the system of selective rewards and encouragements can hardly have failed to reach into science itself, although it was rather later in the century that its full effects were manifest. Presidents of the Royal Society were particularly well placed to exert strong influence. Thus Sir Joseph Banks (PRS, 1778–1820), a wealthy landowner, financed Cook's first voyage, gave weekly receptions at his house in Soho Square for friends to inspect his library and collections, held breakfasts for visiting scientists, returning travellers and explorers and others with whom he wished to converse at length, and was rumoured to have exercised a monopoly control over elections to the Royal Society council.

Private patronage often enabled impecunious authors to publish books that were later to prove of outstanding scientific importance. Thus Sir Hans Sloane (PRS, 1727–40) assisted the zoologist George Edwards to a post as Librarian to the Royal College of Physicians in 1733, thus enabling him both to write and illustrate his impressive and important *Natural History of Uncommon Birds* (1743–51) and *Gleanings of Natural History* (1758–64).[20] Philip, second Earl of Stanhope, financed the mathematical works of the medically trained Robert Simson, sending copies to European learned societies.[21] In England the methods of synthetic geometry, so characteristic of English (as opposed to continental) mathematics in the eighteenth century, were favoured by Simson's republication of ancient Greek mathematical texts.[22] The frequent dedication of works to a noble, or at least wealthy, patron during the eighteenth century, with acknowledgements of help received, indicates the common occurrence of this form of patronage.

Towards the end of the Enlightenment period two examples of private patronage occurred of special importance for the development of science in England. Bryan Higgins (circa 1741–1814) was an Irish physician who settled in London in the 1760s and became part of an influential group of chemical manufacturers and scientific amateurs. Although trying to be self-supporting in his chemical work, he depended on help from several important patrons, amongst whom was George Parker, second Earl of Macclesfield (PRS, 1752–64), an amateur astronomer and mathematician who had assisted another member of the group, William Lewis, in some pioneering researches on platinum.[23] Lewis had also been helped by Hugh Smithson, first Duke of Northumberland, to whom Higgins was indebted for assistance in developing a project for the making of glass. Yet another of his patrons was Field Marshal Conway, an Irish MP and Commander-in-Chief of the Armed Forces.

Patronage was even more important in the case of Joseph Priestley. In 1770 he conceived the idea of publishing a series of volumes on experimental philosophy, to follow his *History and Present State of Electricity*. Anxious to obtain up to £300 for books and apparatus with which to test doubtful experiments he wrote to the Duke of Northumberland, concluding:

> Dr Priestley would think himself greatly honoured, if his Grace the Duke of Northumberland should think this work worthy of his patronage.[24]

He received a loan of books, but no cash, from the Duke, and was compelled to publish over a subscription list. The dedication to the Duke should warn us that such an inscription by itself is not sufficient evidence of actual financial support. It may express thanks for general interest or even optimistic hopes for the future. In fact Priestley's favour with the Duke had unexpected results. Invited to dinner with him in 1772, Priestley was reminded by the production of a bottle of distilled water, intended for use on sea voyages, that he had previously experimented on water exposed to gases from brewery vats, finding it to have a pleasant taste. The conversation over dinner led directly to his invention of 'soda water', originally intended as a drink for sailors on long voyages.

This was not the last of Priestley's adventures in patronage. From 1773 to 1779 he was employed as Librarian by the Earl of Shelburne, later to become Prime Minister in 1782. Aware of Priestley's need for financial assistance (he was then Minister of Mill Hill Chapel, Leeds), but anxious to avoid any suggestion of condescension, Shelburne approached him with tact and delicacy, inviting him to a position that would leave plenty of time for research and writing. All Priestley's conditions were met, and a house was provided at Calne on the Earl's estate in Wiltshire. A salary was offered at two and a half times the value of his ministerial stipend, and he became the friend and travelling companion of his patron. Cataloguing his books and assisting the resident tutor in the education of the Earl's sons left Priestley plenty of time for research. Some of his most important scientific work was done during this period. The arrangement was eventually terminated by Priestley's own choice, apparently because he felt a greater freedom was needed than would be possible in such close association with a leading political figure.

Many other cases of enlightened patronage can be noted: Hutchinson as steward to the Duke of Somerset (p. 63); Rendel the engineer as agent of the Earl of Morley; John Bartram the Quaker botanist and explorer financed by Lord Petre and the Duke of Richmond; and very many others.

There was, of course, a more general sense in which private patronage assisted science (and technology also), and that was in the purchase of books and artifacts. Allen has pointed out[25] how natural history greatly benefited at this time from the production of sumptuous illustrated works which, though intended mainly for display, nevertheless brought a wealth of detailed new knowledge to an audience that would, in its turn, promote further studies in the field.[26] And amongst the industrial entrepreneurs none was more zealous than Josiah Wedgwood in soliciting aristocratic and even royal patronage for his wares.[27]

5.3 VIRTUOSI AND COLLECTORS

Not all science in eighteenth-century England was confined to surrogates acting on behalf of rich patrons. Quite apart from the important if unspectacular development of scientific knowledge amongst humbler classes of persons such as instrument-makers, druggists, engineers and so on (see chapter 6), many other people possessed the right combination of wealth and ability to enable them to treat science as an important leisure-time activity. Thus chemical experiments were sometimes performed in the great houses of the land, most notably perhaps by that eccentric and aristocratic recluse the Honorable Henry Cavendish (1731–1810), nephew of the third Earl of Devonshire and heir to a considerable fortune. More silent (it is said) than a Trappist Monk, intensely shy and forbidding women servants even within his sight on pain of instant dismissal, Cavendish nevertheless managed to attend Royal Society meetings and the soirées of Sir Joseph Banks. And, in his own laboratories, he made major discoveries in electricity, the chemistry of gases and the determination of the density of the earth.[28] Not quite at the other end of the social scale, for he was comfortably off, was Dr Samuel Johnson who in 1772 sent his servant to purchase one ounce of oil of vitriol (sulphuric acid) from a druggist's shop in Temple Bar, for which he paid one penny. Johnson, like many another amateur, conducted 'chymical experiments' on a modest scale, presumably for the fun of doing so.[29]

While many Englishmen had the means to conduct experiments, not all were skilful or rich enough to do so. Their scientific inclinations could be satisfied in less hazardous ways.

Later in the eighteenth century came the Grand Tour, with its opportunities for amateur geologising. But in the first two-thirds of the century an easier method lay near to hand, one that was congruent with the social tendencies of the age: collecting. While some would collect books, antiquities (real or fake) or Sèvres porcelain, others (or even the same people) would seek to amass vast collections of objects of science rather than of art: plants, butterflies, stuffed animals, fossils, minerals, shells – anything that would testify to their enlightened benevolence towards nature. In some cases it is hard to draw the line between this form of activity and patronage. Thus at Bulstrode in Buckinghamshire the Duchess of Portland not only maintained a botanic garden and an enormous natural history collection, but engaged others all over Britain to provide additional specimens. And as her librarian and domestic chaplain she shrewdly employed that eminent botanist the Reverend John Lightfoot. The Duchess appears to have been exceptional, not only in the scale of her operations, but in the genuineness of her scientific interests.

In 1812 the *Gentleman's Magazine* published a manuscript note on collectors during the period 1747–1788 by E. M. da Costa (who had been Secretary to the Royal Society until disgraced by embezzling its funds in 1767). Despite the ignominious end to his career he appears to have kept closely in touch with English collectors over those forty-one years. In his list he included at least twenty-seven people who were scientific collectors in England, five of them Fellows of the Royal Society, nine others being members of the aristocracy or gentry, six engaged in some form of medical practice, and two clerics. Of the thirteen Dutch collectors he mentions, nine were doctors, as were the single examples each from Norway and Spain. What is perhaps most remarkable is that twenty-two of these English collections ended up as sale items, usually in private auctions; the first such sale he recalled in London being that of the collection of Thomas William Jones in 1750.[30]

It is hard to avoid the impression that much of this collecting was to satisfy acquisitive rather than scientific interests.[31] The collecting mania had become fashionable. In France courtiers and noblemen were setting the pace, and in England the aristocracy's lead was to be sanctioned by the examples of

George III and his queen. Indeed, as David Allen points out, the fashion for shell-collecting in France coincided with the shell-like motifs in Rocaille decoration from about 1719, while highly naturalistic floral designs on silks coincided with new interests in botany in the late 1720s.[32] Similarly in the North of England contrasting styles of two silver tea-kettles by Isaac Cookson reflect changing attitudes by collectors; a 1732 kettle was engraved with inter-laced bands and shells, while a 1751 version was covered with flowers, leaves and strapwork.[33]

Fashion, however, was not the only mainspring, though it was undoubtedly a most important one. Da Costa occasionally singles out for special praise those who are more than mere collectors ('virtuosi'), and sought to benefit sciences. Thus of two collectors who lived at the same address in London (Jacob Neilson, a musician and Arthur Pond, an artist) da Costa said 'Pond was only a virtuoso, but Neilson a scientific man', while Sir John Bernard, Bart, is presumably commended for having 'not the least taste for virtuosoship'.[34] Or again, Charles Dubois, who had 'an excellent botanic garden' at the Fair Green at Mitcham in Surrey until his death in the late 1730s, is described as 'a great and celebrated botanist',[35] and Dr Isaac Lawson, Physician-General to the army and an avid collector of fossils, is said to have been the first to introduce the Linnaean system to the learned world.[36]

This last remark is further evidence that acceptance of the Linnaean system was rather a consequence than a cause of the collecting mania. Other benefits to science came from the amassing of really large collections such as that of Sir Hans Sloane which became the nucleus of the British Museum, or the vast 'cabinet of natural history' owned by the French King and kept at the Jardin du Roi in Paris. This was placed in the charge of the naturalist Buffon and gave him the material basis for his epoch-making work on natural history. But for such collections to be of use they had to be as comprehensive as possible and available for long periods of time to those who could use them. Thus they were effectively limited to people with ample leisure or enjoying the benefits of patronage. And even then, of course, the kind of science that could profit was the purely descriptive and classificatory. Since botany and zoology, and to some extent geology, were at that stage of their development they each shared in a limited way in the collecting boom of the early eighteenth century.

The extent to which scientific objectives were perceived as dominant among collectors might be reflected in their social organisation. If, as suggested above, science often played a recessive role in the thinking of most of them it might be deemed unlikely that institutions for scientific debate might emerge, even at local level. In so far as individual acquisitiveness was dominant the sale-room would be more congenial than the salon.

The Royal Society, of course, existed but few collectors were Fellows, and they contributed little to the emerging natural history of the eighteenth century. We therefore have to look for activities at a more local level. In fact, as is well known, the 'Lit. and Phil.' movement did not gain much momentum in England until almost the end of the century, but then, as we shall see, it had quite different origins from the collectors' movement. This movement can be connected, however, with some of the local groups which met towards the beginning of the century.

Perhaps the earliest natural history society in the world was the Temple Coffee House Botanic Club, founded in London around 1689, and having two years later forty members, including Sloane, Dubois, Nehemiah Grew and Petiver. It met for Friday evening discussions and (like the Apothecaries) organised herbalising expeditions into the countryside around London. It was certainly in existence in 1712, though references exist only in the scattered correspondence of its members.[37]

Better records exist for another London botanic club, established by John Martyn, the young son of a City merchant. Meeting on Saturday evenings at the Rainbow Coffee House it had at least twenty-three members, all fairly young. At least half came from fairly well-to-do families (lawyers, barristers, gentry, clergy, etc.). Roughly one-third were destined to become apothecaries, one-third physicians and one-third to practise surgery. So there was a strong vocational emphasis, and this also marked the contemporary Society of Gardeners, which met at a coffee house near the Chelsea Physic Garden. Perhaps in imitation of these clubs there came into being a society of Aurelians, devoted to lepidoptery (and named after a certain butterfly). Its origins are unknown but it survived until its property was destroyed by fire in 1747/8.

By mid-century there is little evidence of locally organised scientific activity in England, a fact that may be compared with the onset of private auctions for scientific collections at exactly this time. The enthusiastic collectors of a previous generation, for whom science had been a genuine inspiration, were dying off, and the spirit of scientific enquiry was being replaced by a collecting mania for its own sake. Like the other branches of science, natural history was experiencing a dramatic decline which may be dated from the 1720s to the 1760s. Allen has pointed out[38] how a youth in mid-century England had to learn ontomology from an octagenarian doyen two generations older. At Oxford and (especially) Cambridge botany was hardly taught and scarcely practised. In this and other areas of science the contrast with the Scottish universities could have created a national scandal, but it did not. Allen may well be right in linking the nearly total stagnation with the lack of leadership and extensive corruption on the national scale from 1724.[39]

The tide began to turn in the 1760s. Changes of personnel at Cambridge and the overdue opening of a new botanic garden heralded better things. And in the next few years various local natural history societies sprang into being amongst working people in Norwich, Eccles, Lichfield and elsewhere. The reasons are obscure. The change in the national political climate may have coincided with the general acceptance of the Linnaean system in England. Later came the slow re-emergence of the Royal Society under the energetic Presidency of Banks. By now the Industrial Revolution was just round the corner and science in general was, even in England, to be perceived in wholly new ways. It is not likely that the *Critical Review* could have announced much earlier than 1763 its confident claim that 'natural history is now, by a kind of national establishment, become the favourite study of the time'.[40]

5.4 SCIENTIFIC TRAINING ON THE CONTINENT

During the Enlightenment some kind of training in natural science became available in many European countries, and, in a few cases, had an importance far beyond that suggested by the scale of its provision. Only the briefest survey is possible here, but even so it must be stressed that, as European cultures varied amongst themselves, so the imparting of natural knowledge took different forms.

We begin with Germany which, in the eighteenth century, was not one country but a large number of independent states with German as a common language. There were over thirty universities, financed by the capitalist State or by the Prince-Bishop.[41] Many were ancient foundations, less changed in their organisation than anywhere else in Europe, and conspicuously displaying their mediaeval origin by their resemblance to the contemporary Guild system for apprentices. In each case the lowest social classes were excluded, the young man would serve his apprenticeship (to Master or Professor) in several places, and a 'masterpiece' of craft or scholarship would be the consummation of his efforts. The later habit of travelling to different universities for research experience owes much to this old tradition of wandering scholars and apprentices. A less desirable outcome was a certain sameness and what Wilfred Farrar has called the production of 'mediocre hard-working pedants totally without originality'.[42] There were a few exceptions, notably Göttingen, founded in 1734 by the British George II who was also Elector of Hanover.[43]

Although the 'masterpiece' system was later to prove effective for training in scientific research, German universities had little to offer in experimental science in the eighteenth century. Almost all there was could be found in the Medical Faculties, mainly anatomical dissection and chemistry. From 1720 to 1750 the number of medical schools with salaried teachers of chemistry quadrupled (from four to sixteen).[44]

From 1740 the programme of reforms under the Empress Maria-Theresa demonstrated convincingly the characteristic features of Enlightenment thinking, especially medical progress and improved education. Thus in 1745 the Dutch physician Gerhard van Swieten was summoned from Leyden to Vienna, there to inaugurate the school of medicine that achieved great distinction in his own lifetime and today is world-renowned.[45] After the Seven Years' War (1756–63) the German states became increasingly prosperous and their autonomy correspondingly at risk from external attack. Provision was therefore made for the training of bureaucrats and army officers, either in law faculties of established universities or in specially created schools. And with an eye to the immense mineral wealth beneath their feet, ruling groups moved quickly to establish new Mining Academies, including Schemnitz (1763),[46] Idria (1763), Freiberg (1765),[47] and Berlin (1770).

The curriculum at a *Berg Akademie* was likely to be narrow, but it did include geology and chemistry. The latter subject had now overcome its earlier stigma of being merely a manipulative art and strongly associated with the dubious practices of alchemy. By the 1770s it had attained a measure of respectability, thanks largely to its successful employment in pharmacy and medicine.

Figure 1 is a graphical summary of data provided by Hufbauer, indicating the number of chemists active (as opposed to being merely alive) in the years specified. The discrepancy between these and the number of posts available (almost all in academic institutions) doubtless reflects the lack of training in earlier periods. Noteworthy is the sharp rise in numbers, of chemists and of chemical laboratories, after 1740 and (especially) after the Seven Years' War. The figure also hints at a temporary down-turn in German activity after 1790. Perhaps the most important fact is that, though the number of chemists more than quadrupled during the century, by 1800 there were still only about thirty in the whole aggregate of the German States. Only subsequent events could demonstrate the importance of such minuscule numbers.

In total contrast to the German States were the Low Countries and especially Holland, which had a long tradition of scientific innovation as the names of Erasmus, Huygens and Leeuwenhoek, and others amply testify. It is no coincidence that, unlike the Lutheran or Catholic States of Germany, Holland was strongly Calvinistic. The University of Leyden

FIGURE 1 Numbers of Chemists and Chemical Laboratories in Germany, 1700–1800.

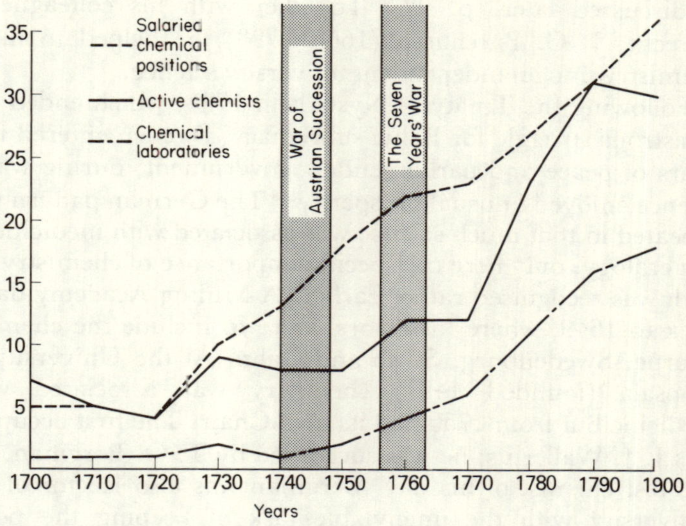

Source: K. Hufbauer, *Hist. Stud. Phys. Sci.*, 1971, 205–231.

had a Protestant foundation (1574) but admitted all comers whether Protestant or not. There was a long tradition of tolerance in Holland; all three competing cosmologies (Aristotelian, Tychonic and Copernican) were taught; Galileo's *Dialogue concerning two new sciences* was published here in 1638 while the author was still under house arrest in Italy; Descartes lived in Holland for over twenty years and Cartesianism was first taught at Utrecht, thereafter gaining wide acceptance in the Netherlands. Despite its speculative character, it was encountered alongside experimental and observational science during this period. W. J. s'Gravesande (1688–1742) from his Chair of Mathematics and Astronomy at Leyden became a leading exponent of Newtonianism in Europe. P. van Musschenbroek (1692–1761) was Professor of Philosophy and Mathematics, and later of Astronomy, at Utrecht before returning to his native Leyden. Both these men received several invitations to other countries, especially Germany, but preferred to teach in Holland. Neither they nor any of their successors achieved, as teachers, the same height of popularity as H. Boerhaave (1668–1738), Professor of Medicine at Leyden. His influence on Scottish University life is discussed later (p. 92). Together with his colleague at Utrecht, J. C. Barchusen (1666–1723), he helped to make chemistry into an independent university science.

Following the Treaty of Nystadt in 1721, which ended her disastrous struggle for Baltic supremacy, Sweden entered fifty years of peace and parliamentary government, during which science enjoyed unusual prosperity.[48] The German pattern was repeated in that much science was associated with medicine or mineralogy, only here the specific importance of chemistry for both was recognised rather earlier. A Mining Academy dated back to 1649, where 'assessors' were to include the chemists Hjarne, Swedenburg, Swab and Gahn. At the University of Uppsala (founded 1497) chemistry was associated with medicine but from 1750 had its own Chair. The first occupant was J. J. Wallerius; he was succeeded by T. O. Bergman, the greatest chemist of his day. Bergman was also Rector of the University with the unenviable task of keeping the peace between science-based academics and those in law and theology. This is itself an indication of an early bid by chemists, pharmacists and others for some kind of academic autonomy. Amongst the achievements of this group of Swedish chemists were the discoveries of the metals cobalt, nickel and platinum, and the great affinity tables of Bergman which led ultimately to a quantified chemistry. Nor must we forget that a Chair of Medicine and Botany was held from 1741 until his death in 1778 by the great Linnaeus himself. Another kind of post existed at the Military Academy, where from 1759, the pioneer of latent heat research, J. C. Wilcke, held a Lectureship in Physics.[49] The Academy of Science was founded in 1739.

A feature of science in pre-revolutionary France was the tradition of public lectures. These were often associated with the Collège Royale and the Jardin du Roi. The Collège, founded in 1530, had Professors in most sciences, including a distinguished succession of chemical practitioners in the Chair of Medicine: E. F. Geoffroy (1709), J. Astruc (1731) and P. J. Malouin (1745). It became the Collège de France in 1793, at the same time as the Jardin du Roi, founded in 1635, was democratised to the Muséum d'Histoire Naturelle de Paris. Intended as an adjunct to the teaching of pharmacy, the Jardin was staffed by a succession of eminent naturalists, including the celebrated Buffon. A long line of chemical professors and demonstrators included Geoffroy, G. F. Roulle, P. J. Maquer and the orator who outclassed them all, A. F. Fourcroy. But despite the popularity of these public lectures it is doubtful if they were effective ways of teaching serious students.

In the provinces, many local académies gave public courses. The Collège de Godrans at Dijon, after being removed from Jesuit control in 1763, gave public courses on physics and chemistry. Even before then it had taught some science, one of its most distinguished pupils being Buffon. In the universities, medical faculties at Paris and Montpellier, especially, taught a considerable amount of chemistry and many of the eminent chemists of the French Enlightenment were medically qualified. The highest standards, however, were usually reached in the military schools and académies where mathematical and scientific training was given, especially in the years just before the Revolution. Finally, for those who could afford it there was private tuition as received by R. A. F. de Réaumur, who contributed extensively to physics and chemistry and came to be one of the most able entomologists of his time.[50]

In these varied ways a tiny minority of European citizens had access to the teaching of science.

5.5 THE SPECIAL CASE OF SCOTLAND

In some previous sections of this chapter the emphasis has been heavily upon English science and its 'organisation'. Some features – notably the collecting mania – were well-nigh universal, but most were not. In the case of Scotland an entirely different picture emerges, the Scottish Enlightenment having its own distinctive character.[51] By about 1750 Scottish science was thriving to a degree unknown before, and was comparable to the best that any other country could offer, with a community of scientists that included such distinguished names as Colin Maclaurin, Joseph Black and William Cullen.

It is tempting, though unwise, to assert that by this time Scottish science was demonstrably better than its English counterpart, but it was certainly different. Before its distinctive characteristics are examined we need briefly to look at the possible reasons for such profoundly different situations north and south of the border.[52]

The most obvious difference between England and Scotland at the beginning of the eighteenth century was economic. Scotland was much the poorer of the two, and indeed the Union of 1707, far from improving the Scottish economy as had been hoped, had actually exerted a depressing effect upon it. There was now direct competition with English traders, and in Edinburgh the disappearance of the Scottish Parliament brought with it a further diminution in trade as the erstwhile Parliamentarians retired to lick their wounds in remote country estates. Meanwhile in Holland, which had close cultural links with Scotland, the economy was strong, and Scottish national pride was the more deeply wounded. Two things immediately followed. One was the need to improve manufactures and trade, and for this science could become the instrument, reflecting 'the drawing power of improvement's ideological imperative'.[53] The other consequence was the channelling of resources into those institutions that were characteristically Scottish. Thus to have all parts of medicine taught in Edinburgh, to improve the University's appeal and to discourage study abroad was not only to boost national pride; it could also make sound economic sense.

A second distinctive feature of Scottish life was of course its Calvinism. Whereas the established Church in England was Episcopalian, with strong Arminian tendencies, the Presbyterian order of the Scottish Church retained in full measure the doctrines of Calvinism and the ethic associated with them. The emphasis on the individual, on the virtues of hard work and self-denial, and on the positive duty to examine and extol the works of the Creator – all these were notes that had become somewhat muted in England. Whereas Newtonianism had been at least in some degree associated with the Latitudinarian Church party and the Whigs, the Scottish divines laboured neither under the necessity of using it as an apologetic tool, nor under the disadvantage of finding it irrelevant to the immediate and earthbound projects that needed their encouragement. And were not the Dutch, a Calvinist nation like themselves, capable of teaching them a thing or two about the road to economic success?

There was also a third feature unique to Scotland and of great importance to its science. In Edinburgh the University, alone among the ancient foundations of Britain, was under municipal control. Its situation in this respect was not unlike that of Polytechnics today. It meant that the Town Council was responsible for the good conduct of University affairs, and especially the appointment of professors, a situation which lasted for over three centuries until abolished by the Universities (Scotland) Act of 1858.[54] This apparently local anomaly had far wider consequences than might have been expected for not only was Edinburgh the intellectual as well as the political capital of Scotland, but also its medical school attracted students from England and beyond, making its University one of the most prosperous in mid-century Europe, second only, perhaps, to Paris.

Thus Scotland differed from England in its dedication to cultural and economic improvements, in its Calvinist ethic and in its University at Edinburgh uniquely tuned to the needs and aspirations of the local and national community. These factors combined to produce certain important developments, most notable among them being the flowering of Edinburgh science in the first half of the eighteenth century.

Although the University had enjoyed several reforms in the seventeenth century, some of its most important changes were inaugurated by William Carstares, Principal from 1703 to 1715. Himself a former student at Utrecht, Carstares was concerned initially with improvements in the training of ministers for the Scottish Kirk. In the event he was to inaugurate a train of events that affected science at least as much as theology. Most important of his innovations was the establishment of specialist professorships in Arts, in place of the system of 'Regents' in which students were taught everything by one man. In 1713 the Town Council, probably manipulated by Carstares, appointed to a new Chair of medicine and chemistry James Crawford, a student of Boerhaave. Although his appointment was not a signal success in terms of teaching or research, in one respect it was noteworthy since it gave professorial status to a teacher of chemistry before even the first text on the subject was printed, viz. Boerhaave's *Elementa Chemiae* (1732).[55] The establishment of a Medical Faculty at the University did not occur until 1726 and then largely through the agency of George Drummond (1687–1766), a member of the Town Council, Treasurer and six times Lord Provost. He was acutely aware of the need for improvement on a national and local scale. His close friend John Monro, a surgeon who had been trained at Leyden, had determined to establish a medical school in Edinburgh on similar lines and, to that end, persuaded the Town Council to appoint his own son, Alexander, as Professor of Anatomy in 1720. By the time the Medical Faculty was established the Town Council had raised the status of professorship in various ways, especially by granting life tenure. In 1726 all five medical professors, led by Alexander Monro, were former students of Boerhaave, and the organisation and curriculum of the Edinburgh Faculty were consciously modelled on their counterpart at Leyden (e.g. in respect of lay control, the central role of the leader who also produced his own text book, and the insistence on scientific before clinical training). As Christie puts it, 'the new Medical School was the high-water mark of that Dutch influence on Scottish higher education initiated by Carstares'.[56] During the eighteenth century no less than forty holders of Scottish academic chairs had studied at Leyden.[57]

Also traceable to Carstares was yet another element in the Dutch tradition inherited by Edinburgh: the attraction to the University of students from other countries, including England. Therein lay much of the unique importance of Edinburgh culture for the development of European science during the Enlightenment.

Today, of course, medical and scientific faculties are usually quite separate, and the development of medical education may have a less obvious relevance to the growth of pure or applied science. In eighteenth-century Edinburgh, medicine was inseparable from chemistry (as, again, at Leyden). Thus we find one of the five Edinburgh Professors, Andrew Plummer, (he died in 1756) teaching James Keir, John Roebuck and

FIGURE 2 H. Boerhaave and Scottish Science.

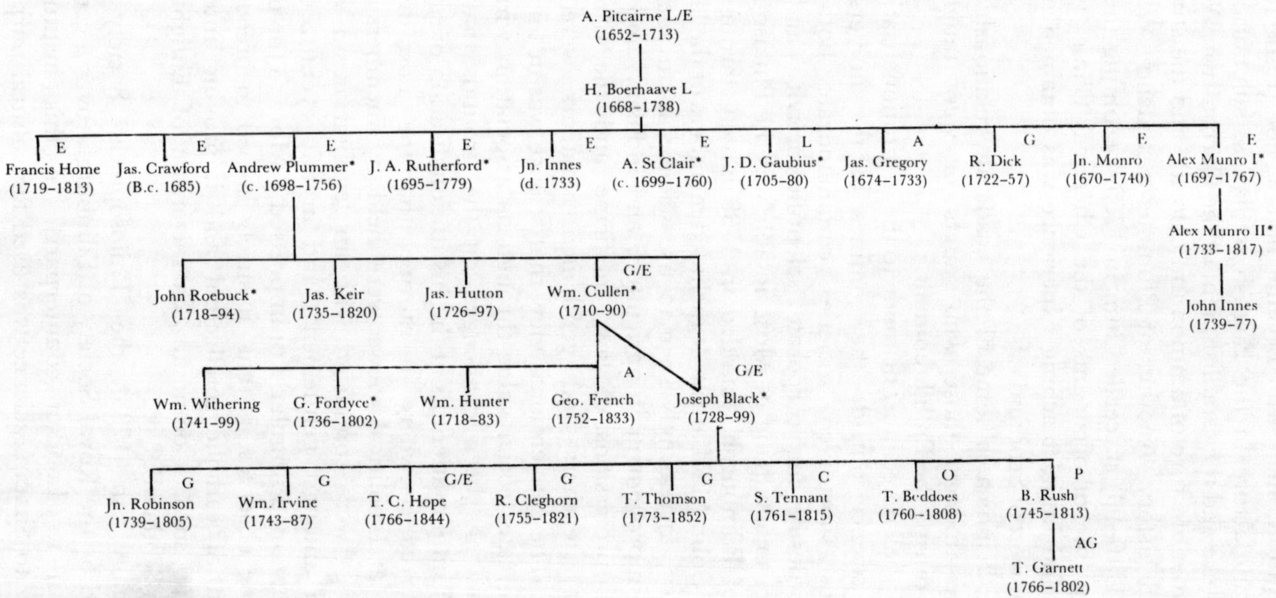

* Membership of Philosophical Society of Edinburgh (ref. 59).

Key to University appointments: A. = Aberdeen, C. = Cambridge, E. = Edinburgh, G. = Glasgow, L. = Leyden, O. = Oxford, P. = Philadelphia, A.G. = Andersonian Inst. Glasgow.

James Hutton, men who became notable for (respectively) alkali manufacture in the English midlands, sulphuric acid manufacture and the establishment of the Carron Iron Works, and the manufacture of ammonium chloride and the study of geological history. Another of Plummer's students, William Cullen (1710–90) succeeded him in 1755, becoming the first of a long line of occupants of the Chemical Chair 'whose training and initial teaching experience was obtained in the industrial west of Scotland'.[58]

Figure 2 indicates some of the academic relationships in chemistry amongst those who taught or were taught in Scotland during the Enlightenment.

Finally, reference must be made to the institutionalisation of Scottish science outside the confines of the universities. Numerous short-lived societies commemorate both the growing desire for corporate scientific activity and the difficulties in achieving it. Most notable was the Philosophical Society of Edinburgh, founded in 1738 largely through the efforts of Colin MacLaurin. It included members of the landed aristocracy, increasingly keen to use science as an agent in national improvement in agriculture and industry, and men from the professions of law, medicine and the Church. University teachers and gentlemen amateurs were also included. The conventional view that science was in this way given a greater social value through its association with the aristocracy has been challenged on the grounds that (for example) similar attempts at the institutionalisation of science had occurred elsewhere in Europe, that the most diverse religious and political views were held by members, that proceedings were conducted with an 'egalitarianism which prevailed amongst gentlemen', and in any event Scottish society remained singularly unimpressed.[59] More to the point, as Christie observes, is that 'it surely helped to breed that contagion of catholic intellectual and improving interests which was such a dominant characteristic of Enlightened society in Scotland'.[60]

Eventually, in 1783, the Philosophical Society was succeeded by the Royal Society of Edinburgh (over a century later than its London counterpart). In this institution, according to Shapin, science was used for cultural purposes, not least on account of its appeal to potential aristocratic patrons. As he put it, 'natural knowledge could be made to serve the purposes of the élite classes',[61] and science was thus seen as not culturally autonomous, but dependent for its organisation upon political, personal and social factors. Along with a strong Tory bias, the Royal Society of Edinburgh welcomed those of other persuasions with sufficient scientific achievements, for example the whig John Playfair. The Society gave an institutionalised base for much research of the highest quality in the century ahead.

Other institutions included the Edinburgh University Chemical Society of 1785; of the fifty-nine original members, fifty-eight had been students of Joseph Black; nineteen were medical graduates (three Scots, three English and thirteen Irish). On this evidence the Leyden tradition was still very active, despite the gradual discrediting of Boerhaave's specific chemical views.[62] Three years earlier a 'Natural History Society of Edinburgh' had been formed as a student initiative. In its first twenty years it had 249 members, of whom about 40 per cent were English and 25 per cent Scottish.[63] An entrance fee of three guineas, an annual subscription of half a guinea and a series of not inconsiderable fines for misdemeanours, confirmed the impression of a much more opulent student membership than is conventionally supposed. Remarks by Kendall[64] about the Student Chemical Society also apply here:[65]

> Our idea of the typical Edinburgh student of the eighteenth century, trudging to and from his home during the Meal Monday Holiday to replenish his sack of oatmeal, obviously needs a little recasting.

Thus in the quality of its teaching, the expectations and organisation of its students, and its application to industry, the science of the Scottish Enlightenment had a quality and emphasis that clearly reflects the different social and cultural conditions of the country.

6. Science in the Early Industrial Revolution

6.1 INTRODUCTION

AS the Enlightenment drew to a close in the final quarter of the eighteenth century so the science that had been its inspiration was to undergo a changing social role. No longer was it to be mainly an ideological weapon in religious or political polemics (though its capacity in this respect was far from exhausted). Nor was it much longer to be a hobby for the rich or a status symbol for social climbers. Its role in society was in fact to change in quite a complicated way, but most obviously in respect of its practical utility. This, of course, is what had been claimed for it ever since Bacon's time, and with some degree of truth. Science had enabled men to navigate more accurately, just as for centuries it had allowed them to calculate the seasons with ever greater precision. Something like scientific thinking had informed those who analysed material used for coinage or who went on herbalising expeditions in the service of medicine. And that boon to mankind, the lightning conductor, was assuredly the high point of applied science in the Enlightenment. But scientific discoveries had chiefly been of use in stimulating further scientific research.

It is possible to argue that the application of science in a wider context had to wait until it had gained in maturity. Science-based technology was therefore a relatively late arrival. Chemistry, the science with the greatest potential in this respect, lacked internal coherence until the very end of the century when the oxygen theory had begun to transform it. It was just at this time that applied science became important, and part of the explanation is likely to lie in external circumstances, including the profound social changes that were beginning to engulf Europe and America. These included the Industrial Revolution (to which the rest of this chapter is devoted) and political revolutions in France and elsewhere which, together with their associated wars, called forth a response described in the following chapter. The cardinal feature of the new situation was that as science reflected social pressures and values so its application proved to be a major instrument in changing the society from which they came.

It is generally accepted that the Industrial Revolution in Britain began in the decades following 1760.[1] Characterised by a mechanisation of industrial processes hitherto performed by individual human beings or by animals, by the harnessing of water-power and steam-power for this purpose, and above all by the introduction of the factory system, the Industrial Revolution involved a complex of social changes of the first magnitude. These included the profoundly important process of industrialisation, the replacement of the old squirearchy by a different form of social control where the labouring masses were effectively governed by their new capitalist masters, by shifts in the balance of intellectual power, and much else besides. It is most generally associated with the rise of a mechanised textile industry with such machines as Kay's flying shuttle (1734), Hargreaves' spinning-jenny (1768), Arkwright's carding machine (1775) and Crompton's spinning mule (1778). At least as important was the steam engine, first applied to a textile factory in 1785 but used for pumping-engines many years before. A third area of vast importance was that of iron manufacture, where replacement of charcoal by coke as a smelting agent dated back to the early eighteenth century, while the 1780s saw the first steam-blown blast furnace and the puddling process for wrought iron. Developments in textiles, steam-power and iron-making interlocked with one another and led not only to an unprecedented growth in each area, but also to the spawning of wholly new techniques, and, of course, to an immense increase in the annual consumption of raw materials, especially coal.

Causes of the Industrial Revolution, especially in Britain, have been debated for many years, and few historians would limit themselves to one explanation. There is no shortage of suggestions: Britain's enjoyment of (relative) peace for a couple of centuries; its unique combination of natural resources (coal, iron ore, timber, etc.); the absence of land frontiers, with all the military implications of that advantage; economic expansion of its overseas markets through colonisation and conquest; the stimulus of a growing population; a specially low rate of interest of government stocks favouring massive investment; and so on. To some extent one's view is likely to be conditioned by ideological preferences and presuppositions. But further detailed examination of local records, particularly, is likely to narrow the possible options.

One factor that is rarely discussed is science.[2] Britain's reputation as the land of Newton is not generally regarded as relevant to the beginnings of the Industrial Revolution. Heroic tales of James Watt and his kettle (with the accompanying exercise in Baconian induction) are no longer worth refuting. Science had little direct influence on the development of technology in the early stages of the Industrial Revolution, even though it was later to become all-important.

The invention and development of the steam engine may have been influenced by science, but only as far as the suggestions of Dennis Papin in the 1690s were dependent upon various gun-powder experiments by Huygens and (more important) on the studies by English mechanical philosophers of the atmosphere and atmospheric pressure. In that context it has been well said that Robert Boyle was not so much 'the father of chemistry' as 'the father of the steam engine'.[3] But the connection between these speculations and the actual engines of Savery, Newcomen and Watt is still problematic. Nor is it very clear how Watt's invention of a separate condenser for the steam engine came from the studies of Black and others at Glasgow University where he worked. On the whole it is safe to say that, following the much later rise of thermodynamics, science owes more to the steam engine than the steam engine owes to science.

However, one part of the Industrial Revolution rather less familiar than textiles, steam power or iron, had a definite dependence on science, and today the only question is the extent of that dependence. This was the infant chemical industry of the eighteenth century.

6.2 THE CHEMICAL REVOLUTION

Developments in the chemical industry in the eighteenth century not only led to many social changes directly but also interlocked with other industries to such an extent that it is no exaggeration to say that, by 1800, this industry was itself an integral part of the Industrial Revolution.[4] That chemical industry itself owed a considerable amount to science[5] can be illustrated by one important example: the case of oil of vitriol, or sulphuric acid.[6]

This highly corrosive, colourless, oily, hygroscopic liquid was part of the legacy of alchemy. It had been made for several centuries by heating the mineral known as 'copperas'.[7] In fact an unusually versatile chemical,[8] it had a variety of uses, few of them well understood. Somehow it was discovered that it can also be made by heating a mixture of sulphur and saltpetre (potassium nitrate) in a glass jar in which water was also present. We now know that an extremely complex sequence of reactions is involved, and there is no possible way in which this process could have been arrived at on the basis of eighteenth-century chemical theory. But in accordance with the wholly empirical nature of technology in the middle of the century it was made on quite a large scale by an English 'quack' doctor, one Joshua Ward, first at Twickenham (1736) and four years later at nearby Richmond. The move may have been due to complaints at the former place because of the unsavoury odours associated with the works. The process was outstandingly successful, reducing the price of sulphuric acid to $\frac{1}{16}$th of its former value.[9] With such a versatile chemical newly available so cheaply it is not surprising that new uses should be found for it, or that others should attempt to emulate and even improve upon the example of Ward.

The most successful manufacturer was John Roebuck (1718–94), a Yorkshireman whose background was highly appropriate for someone who introduced the first elements of a scientific approach to a craft industry.[10] Coming from a dissenting family, he had studied at Northampton Academy, going on to read medicine first at Edinburgh and then at Leyden, places where chemistry was well esteemed and well taught. Settling to a medical practice in Birmingham, he equipped a small laboratory in which he could experiment on metal-refining, an activity much in evidence in that area. Soon Roebuck gave up medicine for chemical manufacture, in partnership with a local business man Samuel Garbett (1717–1805). In 1746 they decided to manufacture sulphuric acid, which was now becoming common as a 'pickling' agent for metals (i.e. for cleaning off oxide films on metal surfaces before further processing). Using the known chemical inertness of lead towards sulphuric acid, Roebuck replaced the fragile glass jars of Ward by 'boxes' made wholly of lead. The acid became even cheaper, and all around were potential users. This was, as the Clows wrote, 'a pivotal point in eighteenth-century economic history'.[11] Three years later a second vitriol works was established by Roebuck and Garbett, at Prestonpans, near Edinburgh. This may have been to avoid infringement of Ward's patent (which covered only England and Wales), but two other explanations suggest themselves. Roebuck may have wished to move nearer Edinburgh where he had learned chemistry and had many scientific contacts. There was also the existence of a vast new market in Scotland.

For some ten years The Board of Trustees for Fisheries, Manufactures and Improvements had concentrated on improving the Scottish linen trade, wishing particularly to eliminate the practice of sending the material to Holland for bleaching. The technique was to boil it alternately with ashes (which contained alkali) and sour milk (whose lactic acid could neutralise the alkali), followed by prolonged exposure to sunlight. Accordingly they subsidised extensive bleach fields over wide areas of Scotland in which the tedious final stage could occur. Very recently the process had been accelerated by replacing the sour milk with dilute sulphuric acid, which compressed into minutes or hours the process which had hitherto taken five days. Roebuck was doubtless aware of this, though the researches of Francis Home, Professor of Materia Medica at Edinburgh, were not published until a few years later.[12] Using improved 'lead-houses' Roebuck was now able to produce enough acid to supply the Scottish need and (as Custom House Returns show) to export to Europe as well. The economic advantages to Scotland were obvious in the same way that attraction of foreign students to Edinburgh was more acceptable than the emigration of Scottish students to Holland (page 92).

After being several times victims of industrial espionage, Roebuck and Garbett tried to obtain a Scottish patent for the process, even appealing to the House of Lords when judgment went against them; but in vain (1771). By now their interest had diversified and, together with five others, they had established the world-famous Carron Iron Works. Roebuck also collaborated with James Watt in the use of the latter's steam engine and in some chemical experiments. All his work was characteristic of the new combination of industrial enterprise and scientific enquiry.

There was a second, and far more dramatic way, in which sulphuric acid was to affect the progress of the Industrial Revolution. In 1774 the Swedish chemist C. W. Scheele discovered a new element, chlorine. This gas was made by the action of sulphuric acid on common salt in the presence of an oxidising agent. In 1787 the French chemist C. L. Berthollet showed that chlorine had a very strong bleaching action on vegetable fibres. This offered a rapid alternative to the tedious method of the bleach-fields.

Within a couple of years of Berthollet's discovery James Watt had been to Paris on patent business, had seen a demonstration of chlorine bleaching and demonstrated it at a bleach-house in Glasgow. His father-in-law, a bleacher, declined to use chlorine gas on account of its dangerous nature. Meanwhile another visitor from Scotland, Patrick Copland, Professor of Natural Philosophy at Aberdeen (Marischal College), while on a continental tour had heard of chlorine bleaching from the chemist de Saussure in Geneva.

Returning home he had the gratification of seeing a local firm, Gordon Barron and Co., put it at once into industrial practice, though using the chlorine in solution rather than as a gas.

James Watt was an admirer and friend of the great chemist Joseph Black, and well versed in modern chemistry. He, like Roebuck and Copland, helped to 'plug in' the burgeoning chemical revolution to the Scottish University system. But he was also a member of the Lunar Society, and sympathetic to political radicalism. His friend Thomas Henry of Manchester soon learned from him about chlorine bleaching, while another Manchester radical, Thomas Cooper, seems to have heard about it from James Watt's son, and established a bleach-works near Bolton in Lancashire.

But it was in Scotland that the greatest effect was felt at first. At exactly this time the country had entered into 'an unprecedented expansion in textile production', increasing production between 1775 and 1812 by no less than a factor of a hundred. This followed a switch from linen to cotton and the collapse of the American tobacco trade with the War of American Independence (1775–77).[13] Such an expansion, in Scotland or in England, would have been totally impossible without chlorine bleaching. Not only did it produce in minutes what would have previously taken up to six months to perform, it also freed large areas of land previously given up as bleach fields.

Various efforts were made to overcome the problems of using the very noxious gas chlorine. One answer, in France, was to absorb it in potash solution (Javelle water). In Scotland, another of Black's contemporaries, Charles Tennant, showed how chlorine could be absorbed by lime to give a readily transportable solid product, 'bleaching powder'. He patented this in 1798 and two years later founded a factory to make it: the great St Rollox Works at Glasgow.

In 1840, sulphuric acid had found yet another use – in the manufacture of superphosphate fertilisers – and, as Justus Liebig said, 'the commercial prosperity of a nation depends upon the amount of sulphuric acid it consumes'.[14] In so far as he was right, the social import of this one product of scientific research must have been, by any account, immense.

There were other science-influenced technological innovations in the chemical field in Scotland at this time. All of them displayed an obvious connection with the prevailing ideology of science-based 'improvement'.

Archibald Cochrane, ninth Earl of Dundonald,[15] in an attempt to make coal-tar for the protection of ships' bottoms against infestation by marine animals, became the first to make effective equipment for the thermal decomposition of coal, setting up a tar-works on his estate at Culross. Another Scot, William Murdoch, an employee of Boulton who had been one of Dundonald's visitors, succeeded in collecting the coal gas and using it for lighting. Later Charles Mackintosh[16] applied the low boiling fraction from coal-tar (naphtha) as a means of waterproofing fabrics. Meanwhile Dundonald had sent an agent (William Losh) to Paris to discover the secret of making synthetic soda. On learning of the Leblanc process he formed a successful partnership in an alkali works on the Tyne. Later he founded the first chair in agriculture anywhere in the world, at Edinburgh in 1790, and in 1795 published a book *Showing the Intimate Connection between Agriculture and Chemistry*. And it is worth observing that utilisation of another by-product of coal decomposition, sal ammoniac, was the work of one much more famous in the history of geology than of industry, James Hutton of Edinburgh.[17]

6.3 INSTITUTIONALISED SCIENCE IN THE INDUSTRIAL REVOLUTION

As we have seen, the thesis that the Industrial Revolution had much to do with science in its early stages is a contentious one. It was part of the received wisdom of many Victorian writers, whose romantic view of the 'heroes of science' was enhanced by the conviction that to them were due many of the triumphs of 'British' technology. Recent authors, writing in a distinctly chillier atmosphere for both science and technology, have been less enthusiastic about such connections, and some have gone so far to the other extreme as to suggest that the practice and promotion of science in the late eighteenth century had been virtually confined to such functions as 'social legitimation', with the applications of science of little consequence. The language may be new but the notion that the Industrial Revolution came about through uneducated empiricism is not. Nowhere has this debunking of 'heroic science' been more strenuously (or solemnly) attempted than in connection with scientific institutions. There is one society, however, which has received considerable scholarly attention and which still remains a prime example of the socialisation of applied science: the Lunar Society of Birmingham.[18]

About the middle of the eighteenth century various improving societies arose in England. Most notable, and most enduring, of these was the Society of Arts founded in 1754 to encourage arts, manufactures and commerce, principally by the offer of prizes. In practice it tended to promote technological progress in a rather limited way: by improving established techniques rather than inventing new ones, and (despite the presence of Stephen Hales among the organisers) by attending to the craft side of industry rather than its potential basis in pure science.[19] Other groups appeared in the provinces with similar objectives, but the first one to pay serious attention to science, as opposed to the useful arts, seems to have been the Lunar Society of Birmingham whose origins may be traced back to 1750.

The Lunar Society was an informal, loosely constituted group of fourteen men, most of whom were to play a major role in shaping the Industrial Revolution. They were:

Matthew Boulton (1728–1809) manufacturer, engineer
Erasmus Darwin (1731–1802) physician, botanist, poet
Thomas Day (1748–89) social reformer, writer
R. L. Edgeworth (1744–1817) barrister, coachmaker
Francis Galton (1753–1832) gunmaker
R. A. Johnson (1745–99) Anglican minister
James Keir (1735–1820) chemical manufacturer
Joseph Priestley (1733–1804) Unitarian minister, chemist
William Small (1734–75) physician
Jonathan Stokes (1755–1831) physician, botanist, chemist
James Watt (1736–1819) engineer
Josiah Wedgwood (1730–95) pottery manufacturer

John Whitehurst (1713–88) clockmaker
William Withering (1741–99) physician botanist, mineralogist

Never keeping records of their activities and only once referred to in print during the Society's period of activity, they were an informal group of friends, linked by a common interest in science and in some cases by family relationships also. The society grew from informal meetings between Darwin, Boulton and Whitehurst, but after 1765 the circle extended to include Small, Wedgwood, Edgeworth and Day, and later on the two Scots, Watt and Keir. A leading role was played by Small, but after his death in 1775 the more casual gatherings were replaced by regularly planned meetings – the nearest they got to formal organisation. The name derived from their practice of meeting on the Monday nearest full moon (thus ensuring a safer journey home at night).

Although only some members were manufacturers on a large scale (notably Boulton, Keir, Watt and Wedgwood), almost all shared an interest in the applications of science, whether to the manufacturing industry or to medicine (as was the case with Darwin, Stokes and Withering). In view of the doubts still heard about the connection between science and technology in the early Industrial Revolution it may be useful to indicate in what senses it could be said that their activities were 'scientific'.

In the first place, they worked with a consciousness of 'universal laws and operations of nature',[20] some of which James Watt prescribed as necessary to be known by a steam engineer;[21] they believed in systematic investigation and inquiry, Wedgwood establishing an 'experimental company', Boulton a research assay office, many of them owning laboratories and most of them making repeated reference to experiments in their notebooks and letters.[22] Conflicting theories were discussed and put to the test, as when Priestley was nearly convinced against his will that his own experiments on water were confirming the theory of Lavoisier.[23] The fact that he, and nearly all his colleagues, were persistently wrong in their adherence to the theory of phlogiston should not blind us to the fact that what they were doing was genuinely science.

Secondly, they were much aware of scientific developments outside their immediate circle. There was systematic reference to the writings of other chemists, botanists and so on; Watt and Keir maintained their intellectual links with Glasgow; non-members like Roebuck and Garbett were in constant touch with them; Boulton and Watt visited Paris in 1786/7 and attended sessions of the Académie des Sciences, meeting Lavoisier, Berthollet and others;[24] Darwin and Withering corresponded extensively with other naturalists in the manner of their time,[25] and distinguished scientific visitors from London and elsewhere such as Banks, Herschel, Smeaton, de Luc, kept them in close touch with scientific developments far beyond the English Midlands.

Thirdly, a good deal of their later work was collective, far removed from the isolated operations of workers in the craft tradition. Together they studied gas-making for balloons, standardisation of weights and measures, heat, colours, chemical changes of many kinds, and together they conducted mineral analysis and (in the early days) worked at astronomy and electricity.[26] Generally they were in substantial agreement on matters scientific, but in the new science of geology controversy reared its head over the vulcanist and plutonist theories of the earth, and in botany and medicine disagreement reached high levels of acrimony. Broadly speaking the 'harder' the science the more united they were, precisely because that was the area where disputes were most susceptible to experimental test and where industrial application was most likely. If science is viewed as a corporate activity, the Lunar Society offers a powerful example.

Fourthly, members of the Lunar Society were recognised outside as practitioners in science. Most were elected FRS, Day, Small and Stokes excepted. Nor was this merely a sign of social acceptance. Johnson, perhaps the least likely of candidates, was admitted in 1788 on the grounds that he was 'well versed in chemistry and other branches of experimental philosophy'.[27] Whatever we may think they were doing, contemporaries had little doubt that it was science.

No amount of learned rhetoric about science as 'social legitimation' can obscure the fact that these men meant what they said. They claimed they were pursuing science for practical ends and knew exactly what they were about. Had they wished to institutionalise certain cultural values they would doubtless have gone about it in a far more efficient way, with a formalised society and plenty of self-serving propaganda. In fact they set out to transform industry by the application of science and, in large measure, they succeeded. The matchless pottery productions at Etruria, the factory at Soho, Birmingham, the ubiquitous Watt steam engine with its overwhelming increase in efficiency on the old Newcomen type, the burgeoning chemical industry, all show their commitment to applied science. Progress would still have been possible without sustained scientific enquiry; and many of their scientific adventures were undertaken for the sheer joy of discovery. But applied science was for three decades a formative influence on their commercial and industrial innovations.

If we can be so confident about the Lunar Society, what of its successors, the 'Lit. and Phil. Societies'? One of the most important of these institutions was the Manchester Literary and Philosophical Society, subject of a number of historical investigations in recent years but of no book-length study.[28] Founded in 1781 in a city where manufacturers generally had a vague awareness of science, the Lit. and Phil. has traditionally been portrayed as the result of a widely perceived need for the dissemination of scientific ideas. Ashton, for example, has in one sentence argued this position for both the Lunar Society and the Manchester Lit. and Phil.[29] The view received powerful support in a paper by Musson and Robinson in 1960[30] who, focusing on the scientific community in Manchester, concluded that there was a substantial number of manufacturers 'passionately interested' in science 'for utilitarian as well as intellectual reasons',[31] and detected in the Literary and Philosophical Society of that city a 'spirit of both utility and idealism', giving a forum for 'an easy exchange of opinion' where 'a manufacturer can consult with half-a-dozen Fellows of the Royal Society among the Manchester members'.[32] They point rightly to the Society's involvement in the formation of first the College of Arts and Science and then the Manchester Academy, and observe that up to 1798, 89 out of the 137 students of that Academy went into industry. They also draw attention to the presence within the Society of several key figures concerned with the making or use of chemicals, including Thomas Henry, FRS, and James Dinwiddie (chemical manufacturers), Thomas Cooper (bleacher) and Charles Taylor (dyer).

Enumerating the other means of scientific instruction (itinerant lecturers, books, free libraries, schooling) they undoubtedly show a connection between some members of the Lit. and Phil. and the applications of science, particularly chemistry. Other early members of the Lit. and Phil. included Peter Ewart the engineer and the science-oriented mill owner Charles Lee. However, one must be cautious.

The Manchester Lit. and Phil. was, even in its early days, much larger than the Lunar Society, and there is a danger of generalising from a small proportion of its membership. If chemistry was the science most likely to be applicable in an area like Manchester, a letter written by James Watt junior offers an illuminating comment. In 1788 young Watt was sent to Manchester to enter the firm of Taylor and Maxwell, fustian manufacturers. The following year he became joint secretary of the Lit. and Phil. and wrote to his father in connection with a new translation of *Crell's Annals*, saying 'Mr Taylor and Mr Henry wish to subscribe for one copy each and Mr Cooper for two, one for himself and one for his partner, Mr Baker. *These are all our chemical men here.*[33] [author's italics]'. Without for one moment doubting the important part played by science in the Society, one may question whether its concern for technology was quite so great as early propagandists wished us to believe. Arnold Thackray clearly does not think so, on the basis of a prosopographic survey of the 588 members up to 1851.[34] Amongst other things he has shown a strong and enduring association between the Society and medicine, with sixty per cent of the founding fathers involved in some kind of medical practice. On this view the growth of science in Manchester was a by-product of the activities of the Lit. and Phil. rather than the major rationale for its formation. But questions must still remain, mainly of a semi-quantitative character. Particularly one would like to know:

1. Since the wealth of data on Manchester science and industry is not always specifically linked with the Lit. and Phil., was that Society truly representative of the wider Mancunian industrial outlook?
2. Given strong evidence that the founders of the Lit. and Phil. were motivated by both technological and more generally cultural considerations, did the latter predominate?

If, as seems probable, the answers to both these questions are in the affirmative, Manchester would offer a striking parallel with Newcastle.

The Literary and Philosophical Society in that city was formed in 1793.[35] Of the fifteen founding members of the Newcastle Lit. and Phil., at least five (thirty-three per cent) can be clearly identified as having medical connections, less than the sixty per cent of Manchester but considerable nevertheless. And at Newcastle there had been since 1787 a Philosophical and Medical Society, from which the Lit. and Phil. drew at least some of its inspiration.[36] As well as the traditional industries of coal-mining and lead-mining in the area, salt had been extracted from sea-water for at least five centuries and sulphuric acid manufacture had begun on a modest scale, but most of Tyneside's important industrial developments appeared after the Lit. and Phil. was formed. They include the Dundonald and Losh alkali works (1797) and the Barnes and Forster Copperas Works (1798), both at Walker. In its first year the Lit. and Phil. received nineteen papers, of which only eight were clearly scientific in theme, four of these bearing some relation to local industry (two on the geology of lead-mines and two on gases in coal-mines). It does not appear that, at first, science was the dominant note of the Lit. and Phil. in Newcastle. Its hopeful intentions to enquire into 'the natural history of coal', to promote a Mines Record Office and to establish a lectureship in science for the benefit of young men in the mining industry in fact achieved little. Support from mine-owners and managers was rarely strong.[37] Nevertheless, as we shall see, it offered a spring-board for the much more extensive institutionalisation of science in the nineteenth century, and from 1803 the monthly meetings were supplemented by lecture-courses which, until 1830, were exclusively devoted to natural science.[38]

In conclusion one important factor must be stressed, common to the Lunar Society and both Literary and Philosophical Societies. That was a tendency towards dissenting religion and radical politics.[39]

One can, of course, overstate this. Thus the Lunar Society, often regarded as a hot-bed of dissent and radicalism, included at least four Anglicans (Boulton, Johnson, Keir and Withering) and was by no means unanimous in its support of the American Revolution (Boulton, Watt, Withering and Edgeworth either keeping a low profile or actually supporting the Government).[40] Most, however, seem to have favoured the French Revolution in its early stages. In the 'Church and King' riots in Birmingham in 1791 the Unitarian and radical Priestley was a prime target, although the cry of 'no philosophers' betokened a popular awareness of links between the Lunar Society and radical politics. Withering's laboratory was damaged and Boulton and Watt (unnecessarily as it happens) armed the workers at their Soho factory.[41]

In Manchester the association of the Lit. and Phil. with Cross Street Unitarian Chapel has often been noted. The Society's first meeting place was a back room of the Chapel. Thomas Henry, a leading light in the early Society,[42] had left the Church of England for Unitarianism some years before. And it is interesting to note that both the Manchester and the Newcastle societies were launched with eloquent rhetoric from Unitarian ministers, Thomas Barnes in Manchester[43] and William Turner in Newcastle.[44] As for political radicals, James Watt Jnr. enthusiastically supported the French Revolution, travelling to Paris in 1792 to congratulate the Jacobin Club on behalf of the Manchester Constitutional Society. He even attempted to mediate in a quarrel between Danton and Robespierre.[45] But, as the Farrars and Scott observed, 'from 1791 to 1794 he lived in revolutionary France, from which he returned rather chastened and with diametrically opposite views'.[46] Then there was Thomas Cooper, another radical, proprietor of the *Manchester Herald,* whose premises were wrecked by a 'Church and King' mob with whom the French Revolution was exceedingly unpopular. Thomas Henry Jnr.,[47] probably a member of the Manchester Constitutional Society, left hurriedly for America to join Priestley.

In Newcastle political radicalism dates back to the political clubs of the 1760s, frequented by no less a person than J. P. Marat who, as a dubiously qualified doctor, combined medical practice and political agitation during his three years' residence from 1770. His *Chains of Slavery*, first published in Newcastle in 1774, was addressed to British electors and purported to catalogue the 'dreadful deeds of despotism' carried out by Princes, urging electoral reforms which were in fact later to be granted. An ingenious experimenter in heat, light and electricity, he was not in the least afraid to contradict Newton. When in 1775 he returned to Newcastle on a visit it was natural for him to call at the new Philosophical Society (not to be confused with either of its similarly named successors). The short-lived association included Thomas Spence, a schoolmaster with pronounced socialist views, whose paper 'Property in Land Everyone's Right' led to his expulsion from the Society.[48] These radical traditions lingered on in Newcastle, despite the arrival in 1796 of over three hundred emigrés from the French Revolution. The following year an Englishman of liberal if not radical outlook settled in Newcastle as a barrister – James Losh,[49] brother of John Losh the alkali manufacturer. In due course James became Recorder of Newcastle, and in 1799 joined the Lit. and Phil., soon becoming its Vice-President. And, like many other members of the Lit. and Phil., he was a Unitarian. That tradition remained significant throughout the nineteenth century for Newcastle science.

Thus in Birmingham, Manchester and Newcastle there was amongst the community of science a strong radical tradition in both politics and religion. The exclusion of precisely these two areas from discussion in the new Lit. and Phil. movement indicates how strongly feelings were known to run and shows not so much their irrelevance to scientific discussion, but the deep relevance they had for the private and sometimes public lives of many of the members. It also demonstrates the belief in the reconciling role, if not of science, at least of intellectual discussion.

Apart from the congruities between the three societies noted above, one may also observe further connections between them. The Newcastle Society had on its membership roll Percival (President of the Manchester Lit. and Phil.), as well as Boulton and Priestley from the Lunar Society.[50] William Turner, the Unitarian Minister, was a friend of Priestley. Many members of the Manchester Society (Walker, Barnes, Cooper, Henry, Percival) were also personal friends of Priestley who, like Watt, also had a son in Manchester.

Of course other groups extended the network. In the Midlands, societies at Leicester, Derby and Nottingham were started in the 1780s by spinners and their associates who had personal links with Priestley and commitment to political and religious dissent.[51] In the West Country the Bath Philosophical Society, founded in 1779, included amongst its members Joseph Priestley, several of his known friends, and the London physician J. C. Lettsom (also in membership at Newcastle).[52] Eclectic in religion, it typically had a fairly high proportion of medical men (thirty-four per cent). Scientifically its focus was on geology, reflecting the local interests in quarrying and building industries.[53] It has recently been argued[54] that the Bath Philosophical Society was one of the most fruitful sources for the work of its best known member and regular contributor: the astronomer William Herschel. He saw his work on the structure of the universe, with its studies of star clusters, nebulae and so on, as a search for natural classes, much in the manner of the naturalists of the Bath Philosophical Society (or for that matter of the Lunar Society or the various Lit. and Phil. Societies). It would be hard to find an area of science apparently more remote from the activities of the industrial chemists and their colleagues, yet in terms of social affiliation and consequent perception of the nature of the scientific enterprise, they are at one.

Further west still the Pneumatic Institution of Bristol (p. 136) acted as a very different kind of centre for science. This was the brain-child of Thomas Beddoes, an Edinburgh-trained physician whose Oxford Readership had to be hastily relinquished in 1792 on account of his indiscreet enthusiasm for the French Revolution. Son-in-law of R. L. Edgeworth, literary collaborator with Keir, correspondent with Erasmus Darwin, Beddoes came to know all the surviving members of the Lunar Society, and obtained the support of most in establishing his experimental hospital, designed to effect miraculous cures with the newly discovered gases. Though very different from the Lit. and Phil. Societies, it became a focus for scientific discussion, and (very nearly) a research institute. Lunar Society members would call there, Watt and Wedgwood sent their sons as patients, literary men with local connections would be frequently in touch (including Southey, Coleridge and Wordsworth's printer Cottle). In 1798 a young Cornishman was taken on as assistant to Beddoes: Humphry Davy.

The Pneumatic Institution gave Davy occasion for his first important researches, into the gas nitrous oxide ('laughing gas'), of which he demonstrated amongst other things its anaesthetic properties. Of even greater consequence was the opportunity afforded to him meet the members and associates of the Lunar Society and to imbibe something of their spirit of dedication to applied science. In 1801 Davy departed for the Royal Institution in London, and shortly afterwards the Pneumatic Institution began to decline. By then Davy had been launched in his London career which, in so many ways, was to exemplify and to extend the corpus of values enshrined in the various provincial science institutes at the end of the eighteenth century.[55]

7. Interlude: Change and Continuity in French Science

7.1 ARISTOCRATIC SCIENCE, 1789–1793[1]

THE Industrial Revolution in Britain was a relatively peaceful transition. The French Revolution was a violent series of convulsions that brought about social changes of the first magnitude. For many years the dominant note in discussions of late eighteenth-century France has been that of discontinuity, though there has been some questioning of that emphasis with a new perception of unbroken threads connecting the Ancien Régime to the Empire.[2] The role of science in these upheavals has been similarly reassessed[3] and a picture is emerging of an increasingly politicised science whose image and function changed considerably from (say) 1789 to 1815. Outwardly innovatory in character, it nevertheless preserved numerous elements from the past, and in these respects partook strongly of the characteristics of the culture within which it was nourished.

The view that the French Revolution precipitated a great crisis for science, that it marked for scientific enquiry as for much else the end of an era, and that there was a decisive break with the past has been well-represented in a certain genre of historical writing since the late 1790s. It has much to commend it. First, there was the abolition in 1791 of the Bureau de Commerce. A feature of the French centralised administration, and dating back to the early years of the century, the Bureau had long involved French scientists (often Academicians) in deciding policy, testing products, assessing inventions and so on. To further its aims the Bureau became involved in the dissemination of scientific knowledge by means of a museum (1782), travelling inspectors (including Gabriel Jars) and books on technical subjects. It had paid employees, including C. L. Berthollet, who invented chlorine bleaching while on its staff. Yet despite its success, the Bureau was one of the first institutional casualties of the French Revolution. Never very popular with that section of the populace on whom it had to enforce its regulations of quality control it fell a victim to the deep mistrust felt by many artisans for the scientists who lived, it seemed, in another world. After the turmoils of 1789 the Bureau found itself subject to mass disobedience, gradual transfer of its various powers to government ministries, and in September 1791, ignominious abolition.[4]

The Bureau was concerned with many matters other than science; not so the Académie Royale des Sciences.[5] While that was functioning science was secure. And in the early years of the Revolution it functioned well, with research in mechanics, dynamical astronomy, electricity and chemistry. It was also grappling with the considerable problem of standardising weights and measures, a process that had symbolic as well as practical significance in the Revolution. An ambitious programme was embarked upon, in conformity with the Académie's wish to impress upon the State its indispensability. In particular it concerned itself with decimalisation of the unit of length. This was defined as 10^{-7} of the earth's quadrant, i.e. the distance over the earth's surface between the north pole and equator. By a combination of triangulation and astronomical observations this length was eventually obtained as 30, 780, 440 Paris feet, the new unit therefore being 3.0780440 Paris feet, the metre. Related to this were units of weight (the gramme) and of currency (the franc), the value of five grammes of silver. Revolutionary governments often have a penchant for decimalisation of currency. It had been discussed in Cromwell's England[6] and adopted by the American Congress in 1786. In France the package of reforms also included the ten-day week (which lasted until 1805). Legal metrication of all weights and measures did not occur until 1840, though even then it was widely ignored in many parts of France.[7]

In the early months of the Revolution many churches were stripped of their bells, and rich emigrés punished by confiscation of their valuables, including silverware. The Académie was requested to help in disposing of the loot. This meant evaluating techniques for assay (chemical analysis) of metals, and development of methods for separating copper from zinc and other metals in bell bronzes.

Additional urgency was given to this task, admirably accomplished by Fourcroy, by the revolutionary wars which began in the spring of 1792. Copper then became in great demand for weaponry.[8] Admittedly publications from Academicians tended to diminish, but this was partly because of diversions on to the war effort and partly because of difficulties with the printer. Despite an effort to maintain political neutrality (or possibly because of it), the Académie began to find itself in troubled waters. In the changing political situation,[9] the Académie found itself succumbing to the bitter factionalism that was sweeping the country at large. Accusation was followed by counter-accusation, and the quarrels were duly reported in the Press. Financial problems compounded the difficulties. As Lavoisier wrote in 1793:

> Time is pressing. A large number of Academicians are suffering, several have already left Paris because their finances no longer permitted them to live here. If help is not on its way, the sciences will slowly fall into a state of decay from which recovery will be difficult.[10]

A report by Condorcet on public education was laid before the Assembly in 1792, but its élitist proposals caused deep polarisation of national opinion and were, for the Académie, the beginning of the end. The Académie itself had always been the incarnation of élitist science. Not even an eleventh hour plea by Lavoisier could save it and, on 8 August 1793, the great Académie Royale des Sciences was closed, along with thirty or more provincial académies of various sorts. More than two years later new institutions were to take their place, but their extinction in 1793 created a discontinuity in cultural history without parallel in modern times. The effect of this discontinuity was intensified by the horrific events of the following two years. It was reinforced by the Jacobin closure of even the engineering schools because of their pre-Revolutionary associations with the army and the monarchy. Yet the view of the Revolutionaries as anti-science vandals determined to sever all links with the past is much too simple, and for several reasons.

Firstly, the problems that led ultimately to the closure of the académies had roots lying deeply in the previous part of the century. The Académie's very exclusiveness was quite sufficient for it to be incompatible with the new egalitarian mood after 1792. And, together with the Bureau with which many of its members were associated, it had attracted great suspicion and even hatred from those artisans whose inventions it failed to sanction or licence or whose scientific expertise it could not approve. If they were anti-scientific (and many of them were) it was partly because of the remote exclusiveness that they felt to be characteristic of the science professed by the Académicians.

Secondly, there was no break in the central administration and its use of scientists to inspect and advise. Often the same people did exactly what they had done all along but in a different context, certainly until 1792. 'There was continuity of the scientists' appraisal of processes because there were persons who had been performing steps in the procedure for many years and they kept on performing them'.[11]

Thirdly, no major attempt was made during these years to reform the French educational system. The fact that school curricula were classical rather than scientific reflects not so much an antipathy towards science as an unwillingness to effect a change in the face of more pressing needs elsewhere.

Finally, two innovations of the early Revolution suggest tolerance, certainly not hostility, towards science. These were greater freedom of the Press and freedom of association. The former made possible the appearance of several new scientific journals which, though they may have undermined the Académie's publications by competition, nevertheless helped to promote science. (The *Annales de Chimie,* founded in 1789, quickly became established in the scientific world and carried many important papers.[12]) The latter facilitated the reconstitution of several pre-Revolutionary societies concerned with the propagation of scientific knowledge. They included the Lycée des Arts, an adult education centre in Paris backed first by aristocrats and then by republicans; the Société Philomathique, a congenial private club for scientific discussions; and the Société d'Histoire Naturelle which more consciously than others attempted to link science and politics.

Thus there was an element of continuity with the science of the Ancien Régime, and Hahn has pointed out the intentionally aristocratic character of French science in the first years of the Revolution. It was some time before science took on any of the character of the Revolution in which it was cradled.

7.2 DEMOCRATIC SCIENCE, 1793–1795

Within less than a month of the closure of the Académie the French Revolution took a new and violent turn. Terror was declared 'the order of the day'. Amid all the horror that followed one event stands out supremely for science, the execution on the 8 May 1794, of Lavoisier, possibly the most distinguished chemist that France has yet produced. His 'crime' was not science, but of being a Farmer General, authorised to collect taxes for the pre-Revolutionary government. In a mass trial of Farmers General, conducted in great haste and with the merest form of legal propriety, they were condemned for 'a conspiracy against the people of France, tending to favour, by all possible means, the success of the enemies of France'. Lavoisier's own scientific reputation was not sufficient to save him; it was rumoured that the presiding judge pronounced 'the Republic has no need of *Savants*; justice must take its course'. Other scientific colleagues also perished in the Revolution, including the astronomer J. S. Bailly (1736–93) and (though not by guillotine) Condorcet himself (1743–94).

Again, however, these events need to be considered in context. The hostility to Lavoisier was undoubtedly directed at his association with the Farmers General, not at his science, and the oft-reported jibe about the Republic having no need of *Savants* has not been substantiated. It appears to have come from the former Abbé H. B. Grégoire, who, together with others after the fall of Robespierre in July 1794, determined to portray the Jacobins' attitude to science in the worst possible light, describing them as systematic 'vandals'.[13] The fate of Lavoisier and his comrades is the clearest possible illustration of the difficulty of separating science and politics during this period. Science was in fact becoming politicised to such an extent that the boundaries between the two areas of human activity became blurred and it was impossible to conceive one without the other. Such politicisation was not new, however, and many would have echoed Condorcet's belief that the laws of natural science should give a model for revolutionary society:[14]

> Science transmits to those who cultivate it something of its noble independence, or it flees from countries subjected to arbitrary rule, or it slowly prepares the revolution which must destroy them . . . in the natural order political enlightenment follows in its wake.

Numerous French scientists were noted for their strongly radical political views: Carnot, Prieur and Guyton de Morveau were all members of the Committee of Public Safety, and Fourcroy, though rather a political chameleon, was a member of the new National Convention (as was Carnot). Only Priestley had comparable views amongst the major British scientists, although, as we have seen (Chapter 6) they were shared to some degree by numerous minor figures. But the politicisation demanded in the years 1793–5 went much further than the mere adherence by a scientist to a general political ideology. It meant an order of priorities: politics before science. 'Correct' political views were more important than scientific rectitude, hatred of monarchs was of greater value than hatred of phlogiston. How else could the execution of Lavoisier be satisfactorily explained? Even in 1790 it had been recognised that 'one must be a citizen before being a naturalist'.[15]

Science was now to be pursued in a manner compatible with

revolutionary sentiments. That was why the élitist Académie had to go, and that was the rationale for a scheme to replace it, drawn up by Fourcroy and Romme. Not only was the former exclusivism dispersed, but so also was the possibility of ministerial patronage. In its place was a strongly centralised committee system.

In fact this ideologically respectable scheme proved to be, as Hahn has put it 'a miserable failure', in part because of the removal of that Académie which had played a central role directing and stimulating French science for over a hundred years. The new arrangement 'was to shatter the bonds that had previously countered the centrifugal tendencies of specialisation'.[16] That is at least part of the explanation of why Lavoisier was left to his fate by fellow scientists: he was simply no longer their colleague. The community of science was hopelessly fragmented. The 'free societies', such as the reconstituted Société Philomathique, were no substitute for the great Académie, and they were half-heartedly supported. Sometimes not even a quorum turned up for meetings. A politicised science was making other demands, not least in aiding the French war effort.

The direction of scientists to militarily useful work had just predated the fall of the Académie, when Guyton appointed four citizens, including Berthollet and Fourcroy, to research into new techniques for defence. This 'congress of scientists' produced spectacular results in steel-making, organising munitions production and continuing Fourcroy's work on extracting copper from church bells. Most notable of all perhaps, was their production of saltpetre.

Saltpetre (nitre, potassium nitrate), is one of the constituents of gunpowder, together with charcoal and sulphur. Before the Battle of Saratoga (1777) most of the gunpowder used in America was exported from France.[17] The saltpetre normally came from India, but now, in August 1793, the country had less than twenty-five per cent of its total needs and no guarantee that fresh supplies would get through by sea. The Committee of Public Safety turned to the chemists to provide a solution, and this they did within a matter of weeks.

Nitrates result from microbiological oxidation of nitrogenous organic material in the soil and may be isolated in crude form from soil, near stables, barns and even cellars. Hence it was to those sources that the populace was required to go for its nitrates. The action proposed caught the attention of Thomas Carlyle who described it in characteristically colourful terms:

> See! The Citoyens, with up-shoved *bonnet rouge*, or with doffed bonnet, and hair toil-wetted; digging fiercely, each in his own cellar, for saltpetre. The Earth heap rises at every door; the Citoyennes with hod and bucket carrying it up; the Citoyens' pith in every muscle, shovelling and digging: for life and saltpetre. Dig, my *braves*; and right well speed ye! What of saltpetre is essential the Republic shall not want.[18]

Eventually production rose to five hundred pounds per day. Unfortunately Carlyle's rhetoric was better than his chemistry. The product obtained in the neighbourhood of old cement and mortar, as in cellars, is not saltpetre but calcium nitrate. This requires to be converted into saltpetre by the action of potassium carbonate, another product usually imported but capable of being extracted from the ash from burning wood and various plants. But this posed another problem, for this 'potash' was also required for different purposes (especially the manufacture of soap). For this, however, an alternative was soda (sodium carbonate). Accordingly the Committee proposed:

1. The more systematic and efficient burning of trees and shrubs which produce potash;
2. Development of alternative methods for the manufacture of soda.

Each of these was put into practice, the former providing as a by-product extra charcoal which could be used for gunpowder. Figure 3 shows the products and by-products of the French manufacture of gunpowder.

The idea of making soda (sodium carbonate) from salt (sodium chloride) was not new, various processes having been tried in France, England, Scotland and elsewhere. As far back

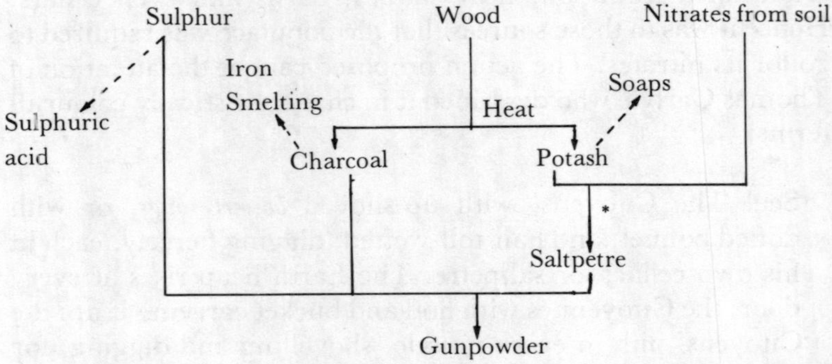

----→other uses

*The reaction is

$$Ca(NO_3)_2 + K_2CO_3 \longrightarrow CaCO_3 + 2KNO_3$$

Saltpetre

FIGURE 3 French Manufacture of Gunpowder.

as 1782 the Académie had offered a prize of 12,000L for such a process, but no award had been made. From about 1784 to 1788 a new and promising process was being developed in Paris by Nicholas Leblanc, formerly surgeon to the Duke of Orléans. This involved heating salt with sulphuric acid to yield sodium sulphate (saltcake), and then heating this in a furnace with a mixture of limestone and coke. From the unpromising looking product soda could be extracted with water and then crystallised out. In 1791 Leblanc lodged a patent application and began to work the process at a site in Paris, with the aid of a loan from the Duke. The patent was unnoticed and filed away, but was exhumed in the crisis of 1793 and the works taken over by the state in January 1794. (The land belonged to the Duke, but as he was executed in 1793, it was confiscated and thus became state property in any case.) Leblanc received his factory back in 1801 but, deeply in debt and having gained nothing from his patent, he shot himself in 1806, a belated casualty of the French mode of politicising science. But his process became the basis of the world-wide heavy chemicals industry for over a century.[19]

In the matter of saltpetre manufacture the resourcefulness of the French Government and its scientist employees stands in marked contrast to the British experience. Britain had long

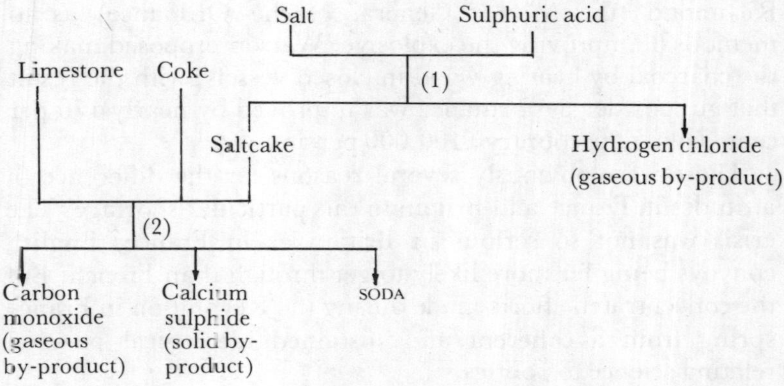

The reactions are:

(1) $2NaCl + H_2SO_4 \longrightarrow Na_2SO_4 + 2HCl$

(2) $Na_2SO_4 + CaCO_3 + 4C \longrightarrow Na_2CO_3 + CaS + 4CO$

FIGURE 4 The Leblanc Process.

relied upon the East India Company for its saltpetre supplies, but for five years that Company had been in arrears with its deliveries, and, before the war with France was a month old, alarming shortages were becoming evident. When the Company's saltpetre warehouse in London was destroyed by fire in July 1794 (possibly by sabotage) it was probable that nearly all the country's saltpetre went up in flames. In this matter Britain was obviously caught completely unprepared. Measures to deal with the situation were chiefly to seek purchases abroad, especially from Sweden. Otherwise action was limited to declining a request from the Irish Government for one thousand barrels of fine powder, and 'by stopping all salutes being fired at the present moment and during the war'.[20]

British failure to seek a scientific solution to the saltpetre shortage is the more surprising since a laboratory had existed at Woolwich since 1747. Moreover, the quality of gunpowder between the American and French Revolutions had been greatly improved by the application of scientific research. Richard Watson, Bishop of Llandaff and former professor of chemistry at Cambridge, had been consulted by the Duke of

Richmond (the Master General of the Ordnance) as to methods of improving the explosive. Watson proposed making the charcoal by heating wood in closed vessels, with the result that gunpowder performance was improved by nearly 170 per cent, saving the country £100,000 per annum.[21]

There were obviously several reasons for the difference in attitudes in France and Britain to this particular shortage. The crisis was not so serious in Britain as in France, English convoys being far more likely to get through than French. But the concentrated efforts made during the Revolution in France spring from a coherent and sustained ideological position relating science to politics.

An even more striking contrast was an American attempt to mobilise the nation's saltpetre resources. In 1798 the Chemical Society of Philadelphia (founded in 1792) established a Nitre Committee, inserting a notice in newspapers throughout the United States seeking information about its manufacture in different countries. Citizens were invited 'to forward it to them (post paid)'. This considerable exercise of private initiative does not seem to have borne much fruit. In the absence of a nationally perceived crisis in saltpetre, and without official support, this is hardly surprising.[22]

Apart from the direction of scientific manpower the most notable and enduring features of 'democratic science' in France were in the field of education.[23] At the secondary level little had changed until 1795, when proposals were canvassed for central schools with some science and modern history and languages replacing the hitherto classical curriculum. Science education with a technical emphasis had existed in the Ancien Régime, mainly in colleges of engineering, and they continued to function through the early years of the Revolution, though with very low productivity. Public lectures in chemistry had been popular, notably those of Guyton de Morveau in Dijon and those by the powerful orator Fourcroy in Paris at the Jardin du Roi. Courses of this kind continued until the Terror when, according to newspaper advertisements, their numbers dropped dramatically, due undoubtedly to the 'pervasive sense of demoralisation that presided over the scientific community'.[24] A new urgency was given to technical education by the outbreak of war and early in 1794 four 'Revolutionary courses' were launched by the Committee of Public Safety to give intensive instruction in such matters as saltpetre manufacture and production of guns. They were shortly followed by similarly intensive though wholly unsuccessful crash courses in teacher training. In all these activities ex-Académicians like Fourcroy, Guyton and Berthollet figure prominently. At the same time a shortage of doctors in military service led to the resurrection of medical schools in Paris, Strasbourg and Montpellier. At Paris, Fourcroy taught chemistry to students whose qualifications had to include some knowledge of science and a hatred of tyrants.[25]

Most important of all the educational innovations was the foundation in December 1794 of what was later to be called the École Polytechnique.[26] Intended primarily for engineers, it nevertheless included a surprisingly large amount of chemistry at first – about a quarter of the students' time. Again the teachers included Fourcroy, Guyton and Berthollet. The prominence of chemical teaching reflected the authorities' confidence in the political trustworthiness of these men, and also the proven utility of their science in such matters as metal extraction and alkali manufacture. Regular visits were arranged to factories, and several small industrial plants were installed within the college. Most remarkable of all, however, was the inclusion of systematic laboratory training. Hitherto this had been routinely available to only one large group of chemistry students anywhere in Europe (at the Kayserliche Köngliche Bergwesens-Akademie at Schemnitz in Hungary, founded in 1770 to service the expanding metallurgical industries of the area[27]).

7.3 BUREAUCRATIC SCIENCE, 1795–1799

Towards the end of 1794 the worst of the Terror was over. Robespierre had been executed in July, the Jacobin Club was closed in November. In 1795 peace was concluded with Prussia, Holland and Spain and in August of that year a new constitution was established which meant that the old Convention was dissolved and replaced by the Directory of Five, with two legislative assemblies. That year also saw a return of institutionalised official science. In place of the Académie there emerged the Institut de France, one manifestation of the Convention's determination to turn its back on the immediate past and to reconstitute broken links with the Enlightenment. This was a National Institute of Arts and Sciences, differing from the old Académie in several ways. It was not simply scientific but had three classes, the first of which only was for science, the second for moral and political science and the third for literature and the fine arts. Yet all three were now to be consulted regularly by the Government.

The organisation was strictly meritocratic rather than aristocratic. But it was now no longer alone, and research and consultation were being carried out by other centres of excellence, such as the École Polytechnique, the Muséum d'Histoire Naturelle, and so on. However, membership of the first class was ardently sought by budding young scientists. Salaries were to be paid for the first time, inherited wealth no longer being the guarantee of admission.

It can be argued that one reason for the Institut's establishment was a reaction against the notion of Jacobin vandalism towards science, systematically orchestrated by those who wished to establish a successor to the Académie.[28] The plain fact is that very many voices were raised in the mid-1790s expressing sentiments similar to those of Fourcroy:

> Our last tyrant, who himself knew nothing, who displayed gross ignorance, who collected evidence against some of his colleagues partial to knowledge and science, and who would have sent them to the scaffolds . . . looked upon learned men with suspicion, fury and envy . . . because he knew they would not bow down to him.[29]

Granted that draconian action against science or scientists was only sporadic, science was certainly a part of the culture which was ruthlessly lacerated during the Jacobin ascendancy. It is hardly surprising that passions should be so high, and difficult to see how else one could account for 'the eclipse French cultural life had suffered during the Terror'.[30] An argument used for a new order does not of necessity rest on false premises. There were, of course, other reasons for founding the Institut, not least the widespread desire for a return to Enlightenment values and the now demonstrable need for Government sanction of cultural activity. The characteristics of this new institution, with its central control, salaried officials, meritocratic basis and role as Government adviser strongly justify Hahn's description of the science of this period as 'bureaucratic'. Indeed this remains substantially true in the modern state of today.

If it is asked how the government viewed the science over which it now exerted some control, it has to be admitted that the idea of 'science for its own sake' was largely, though not completely, absent. There is little evidence that the early Revolutionary fear of an autonomous republic of science had vanished in the era of the Institut. The Institut was not the Académie writ large. Emphasis still tended to be placed on the utilitarian aspects of science, and projects such as the standardisation of weights and measures continued to progress. In 1798 an International Congress was convened in Paris to ratify the decisions made. It was probably the first international scientific congress ever held (though France's warring enemies were naturally excluded) and invitations were sent, not by the Institut, but by the Minister of Foreign Affairs, Talleyrand. Clearly France's position in the eyes of the international community would be enhanced by such a prestigious event, and that is certainly how it was seen abroad. 'The Institut and the French Government have given a grand and noble example to the world' wrote a delegate from Holland.[31] Science was becoming a symbol of cultural superiority.

To see science in that light was to apply pre-revolutionary values in a post-revolutionary political situation. The Enlightenment view of science as a paradigm of progress was also important in the very structure of the Institut itself. Though science was now only one of three classes it was (thanks to Fourcroy's machinations) much the largest of the three. And the whole organisation was planned in the light of experience with science alone, i.e. in the old Académie.

At the same time as its utility was becoming widely recognised in late revolutionary France, science was undergoing a further transformation of its role within the total culture. As Maurice Crosland has stressed a 'more absolute differentiation' between arts and sciences occurred within the Revolution.[32] Partly this was because of the applications of science generated by war, but it seems that the structure of the Institut itself was also responsible. Placing together arts, literature and sciences in one institution actually heightened the sense of differentiation between them. But more than that, Fourcroy's plan involved gathering within the first class all those 'useful arts' that were science-dependent. In this way the first class became exceptionally large, and pure and applied science were culturally insulated from the other areas of intellectual life. There was no intermediate category of 'useful arts' to fudge the clear-cut issue.[33]

This stratagem mirrored the growing differentiation of science within French culture as a whole. It doubtless helped to increase that differentiation further still. Science was now becoming something like a profession, thanks in part to the proliferation of specialist scientific education, especially in the École Polytechnique. To quote Crosland again, 'the Revolution marked both a quantitative and qualitative change in the position of science. . . . Science not only had a significantly greater institutional support, it was now a nationally recognised activity inviting talented young men. . . . Science could now provide not only an intellectual challenge but a career. A young man who acquired a scientific education by availing himself of the free courses of lectures given in Paris could aspire to fame and fortune in this newly opened field'.[34]

Associated with this 'professionalism' was a greater specialisation of activity which meant, strangely enough, a weakening of the Institut's central position, as small groups of scientists formed semi-independent cells within the École Polytechnique, Museums, Observatory and other centres where their specialism could more effectively find expression. At the same time the Institut began to lose some of its adjudicatory functions (e.g. in patent awards) as these were hived off to central government departments. And increasingly scientists felt frustration from their parity with the other classes, since each had to have an equal space in the *Mémoires*. Scientific papers could therefore be delayed for years while poets and other literati were waiting for the urge to blossom into print and thus fill up their allocation of space. Until this absurdity was rectified (in 1798) scientists naturally looked elsewhere to publish and, once having got into the habit of writing for *Annales de Chimie*, *Journal de Physique* or one of their contemporaries, would continue to do so. But as the projector of the national scientific image the Institut remained supreme. Its links with the State became closer than ever, and not only scientists but successful men of affairs thronged to join it. One of the latter, elected in 1797, was a young and successful military man who, in two short years, was to gain a position from which he personally would exert a profound influence not only on the Institut but upon science in France and upon the course of world history. His name was Napoleon Bonaparte.

7.4 IMPERIAL SCIENCE, 1799–1815

As the eighteenth century drew to a close the Republic, bankrupt and exhausted with war and revolution, slipped quietly into the hands of the only man who could fill the political vacuum. The victorious general recently returned from Italy and Egypt, Napoleon Bonaparte, carried out a coup d'état in November 1799. Less than five years later, following the discovery of a Royalist plot, he crowned himself Emperor of France. Long before this, however, his style (to say nothing of his ambitions) had been imperial and, until his defeat at Waterloo in 1815 and his banishment to St Helena, all French culture, including science,[35] reflected in some measure the cult of the Emperor.

When Charles Babbage wished to attack the Royal Society and to condemn English science by unfavourable comparisons, his somewhat jaundiced eye fell upon the France of Napoleon:

> The *Mecanique Coeleste*, and the *Théorie Analytique des Probabilités*, were both dedicated, by Laplace, to Napoleon. During the reign of that extraordinary man, the triumphs of France were as eminent in Science as they were splendid in arms. May the institutions which trained and rewarded her philosophers be permanent as the benefits they have conferred upon mankind![36]

This favourable view of Napoleon's personal influence has been widely echoed. Certainly the years 1799–1815 saw much spectacular scientific progress. In 1808 Gay-Lussac discovered his law of gaseous volumes, and, jointly with Thenard, isolated potassium and boron. In optics E. L. Malus discovered the polarisation of light (1808) and Arago went on to discover optical activity (1811). Much useful work was done on heat, not least the measurement of specific heats by F. Delaroche and J. E. Bérard in 1813. Journals continued to flourish, and important books appeared, like the two by Laplace which Babbage mentioned and Berthollet's introduction of the concept of chemical equilibrium in his *Essai de Statique Chimique* of 1803.

The extent to which Napoleon can claim credit for the progress of science is debatable. Without doubt he encouraged application of science to agriculture and industry. He seems to have been specially anxious to support the new science of electricity, welcoming to Paris Alessandro Volta, who invented the first electric battery, and installed a large voltaic pile in the École Polytechnique. Napoleon persuaded the Institut to award a prize of 3000 francs for the best electrical research each year and was probably pained when the first was won by an Englishman, Humphry Davy. Above all, he promoted both directly and by example a complex system of patronage. One writer has observed:[37]

> Unlike the case of the liberal state which was to succeed it, patronage was the dynamic of Napoleonic government.

It was much more important than institutions in stabilising the life of science.

A classic case of patronage is the private club which came to be called the Society of Arcueil.[38] In 1801, Berthollet acquired a country house at Arcueil near Paris, equipping it with laboratories and library to be a centre for research in physics and chemistry. Five years later his friend Laplace bought an adjoining property, and the two men became leaders of a group which included Gay-Lussac, Humboldt, Thenard, de Candolle, Descotils and a few others. They provided laboratory facilities, fortnightly discussion sessions on Sunday, and (from 1807) a journal for the publication of research. In these ways they exerted a powerful patronage on the members of the group and materially assisted their careers. They were able to do this because, as Senators, they received substantial stipends, and this privilege in turn they derived from Napoleon. Berthollet had been a companion on his Egyptian tour, and 'was almost in the position of a royal favourite.'[39]

Nor was Arcueil the only channel of patronage. This was the age of the French salon in which wealthy and influential ladies would open their drawing-rooms to men of science and letters. The Society of Auteuil, concerned with life sciences, began in the home of Madame Helvetius, friend of Napoleon and widow of the writer C. A. Helvetius.[40] Individuals such as Biot were indebted to the mathematicians Lacroix, Lagrange and Laplace for advice, support in election to the Institut, teaching position, access to publishers and so on.[41] A patron, for his part, needed institutional, social and family contacts[42] and, because he also needed state employment, in the end he needed Imperial favour.

The view that Napoleon was less than a whole-hearted patron of science has been voiced by several writers, including Robert Fox. He points out that 'considerations of utility or national prestige were never far from Napoleon's mind' and that his measures 'showed little sympathy for science as an intellectual activity'.[43] To Napoleon must be ascribed at least some of the blame for the run-down of the École Polytechnique, culminating in its 'militarisation' in 1804, when students had to don military uniform and live in barracks. Admittedly the decline had set in before Napoleon's coup d'état of 1799, and in the ten years from 1794 the chemical content of the curriculum had dropped from twenty-two to two per cent.[44] Again, Napoleon's own educational reforms had to wait until 1802, and then they were far from adequate. The view that they went on to play a major role in science education has been defended by R. R. Maras[45] and attacked by Pearce Williams.[46] Robert Fox has argued that the successes of French science in the Napoleonic period resulted from individual initiative, rather than government sponsorship, and that the subsequent decline of French science can be attributed to a lack of such private initiative.[47] Evidence for such individual motivation is strong. The distinction achieved by many young physicists (Biot, Poisson, Dulong, Arago, and Petit among others) can hardly have derived from their previous training at the École Polytechnique, for their teacher was J.-H. Hassenfratz who was, on all accounts, a thoroughly incompetent Professor of Physics from 1794 to 1815.[48] But chiefly it was in the personal dedication of Laplace and Berthollet that the resurgence of scientific enquiry pursued for its own ends and without direction from the State could be seen. They both were, paradoxically, protégés of Napoleon; they symbolised a reaction to the old polarisation between utilitarian pursuits sanctioned and enlightened by science, and the higher flights of untramelled scientific enquiry pursued for quite different ends. It is noteworthy that, of the former Académicians, Berthollet was one of the least politically committed, while Laplace was one of those most able to 'trim' to the prevailing political philosophy.

In a finely judged analysis of the relations between French industry and science, C. C. Gillispie has distinguished between two kinds of science.[49] Taking science as the pursuit of knowledge for its own sake he argues that, in the supposed application of science during the war to improve armaments, make saltpetre and so on, 'science was only exploited. What was applied was scientists'.[50] He argues that the kind of scientific approach being employed was that of the natural historians, to classify, arrange and explain technological processes. 'The inspiration was algebra. But the model was botany.'[51] Very few of the French technical developments, including the Leblanc process, came about from sustained scientific enquiry into the underlying laws of matter or motion.[52] Consistent with this view is the additional explanation advanced by Langins[53] for the deceleration of science in the École Polytechnique; she attributes this not only to the lack of funds but also to the emphasis in Fourcroy's teaching of chemistry. His attitude, typical of the eighteenth century, was essentially static, descriptive and explanatory. He did not, in the manner of Lavoisier, seek out new laws with great predictive power. He worked much more in the tradition of natural history, a subject on which he also lectured occasionally at the École.

Given a reaction to revolutionary doctrines after 1799, an income from the State so large as to mean virtual financial independence, and the added prestige of age and seniority, it is not surprising that Berthollet and Laplace were able to return to the spirit of free scientific enquiry associated with the high noon of the Académie. In that sense, together with the fact that science could be a matter of national pride, it is not inappropriate to style their activities as 'imperial science'. But the phrase may seem to have a deeper significance if one enquires into the nature of their scientific enterprise and its motivation.

Laplace and Berthollet were both committed to a programme of Newtonian research. Some years later the physicist Arago observed of Newton's discovery of universal gravitation that no Frenchman could reflect without an aching heart on the small participation of his own country in that memorable achievement.[54] Conceivably patriotic motives of this kind may have had their place in promoting a Newtonian programme of research. What is more certain is that both Laplace and Berthollet sought to bring, respectively, the whole of physics and chemistry within the jurisdiction of Newtonian law. In that sense their 'imperialism' was of course more ideological than military.

In Laplacian physics the assumption of forces acting at a distance was complemented by that of imponderable fluids such as caloric (heat), light, electricity and so on. Whereas Newton had brought the whole of celestial motion, at least in principle, within the scope of one gravitational law, his ideas about the same forces operating at microscopic levels had remained in the realms of speculation. Laplace was to preside over a comprehensive programme of research aimed at extending Newtonian principles to all parts of physics. His patronage was well rewarded with the discoveries by Arago and Malus, together with Biot's work on optical refraction. Berthollet's programme was arguably less successful, but he attempted to treat chemical affinity in the same way that Newton treated gravitation (without necessarily identifying them): as dependent on mass and on distance between attracting particles. These views were enshrined in his *Essai* of 1803, where the influence of mass on the position of chemical equilibria is clearly shown, and where variability in the composition of certain chemical compounds was given credence. His work was continued by his pupils Gay-Lussac and Dulong.

During the Napoleonic period Laplacian physics was not universally accepted, and it came in for some criticism from the mathematician J. B. J. Fourier who, though a protégé of Napoleon, was far removed from the Arcueil circle by virtue of his Prefectship of Isère. His mathematical work on heat conduction in solids (1807 onwards) owed nothing to Laplace and little to Newton. After 1815 Fourier's revolt against the Laplacian programme was joined by Dulong, Arago and (especially) Fresnel, whose support of the wave-theory of light was totally subversive of Newtonian concepts of imponderable fluids. Others joined them, and, as peace-time relations with England were resumed, French science became more strongly exposed to the Daltonian Atomic Theory, which, once accepted, eroded the foundations of Berthollet's thesis.[55]

In fact the end of the Empire marked at a deep level a major discontinuity in French physical science, even though institutional changes were modest and in other respects science appeared to suffer little from the restoration of the monarchy. A series of apparently unconnected discoveries, such as electromagnetic induction by Oersted and the Law of Atomic Heats by Dulong and Petit, a sudden drop in income for Napoleon's protégés Berthollet and Laplace and a growing antagonism to their stranglehold on French science, all conspired to bring about a dramatic decline in their influence. By about 1825 French science had entered into a long period of decline for which some responsibility may be laid at the doors of the Laplacian domination of Napoleonic science. As Robert Fox has well observed of this most distinguished school of French science:

> As an illustration of the power of 'totalitarianism' in science, it is perhaps without equal, and . . . it displays the dark side of totalitarianism. For in the perpetuation of ideas that were kept immune from rigorous criticism at a time when reappraisal could have been beneficial there lay the dangers of the orthodoxy that Laplace and his followers tried to impose on French science.[56]

Thus the development of this kind of science under the first Empire demonstrates factors of supreme importance and simplicity, all too easily lost sight of in a welter of arguments about the relation between science and state in modern society. Given genius, financial autonomy and powerful personal ambition, science can indeed flourish and rise to great heights. But for it to become part of a continuing and enduring programme of scientific research, there has to be freedom from all kinds of totalitarian repression.

8. Radical Science in Britain, 1790–1830

IT was the misfortune of Dr Thomas Beddoes to make enemies nearly as easily as he made friends. In November, 1792, a confidential note was passed to Mr Isaac Hawkins Browne, of Badgere near Shiffnall, from the permanent Under Secretary at the Home Office, Sir Evan Nepean. It reports that Beddoes was sowing sedition in Browne's area and seeks information.[1] A few months earlier a question had arisen of Beddoes' promotion at Oxford from his self-styled Readership to an official Chair. A local JP, C. Willoughby, of Baldon House, did not consider Beddoes to be 'deserving of his Majesty's bounty'; although a good chemist 'he is *a most violent Democrat* and . . . he takes great pains to seduce young men to the same political principles with himself'.[2] With political unpopularity matched by declining student numbers (and therefore fees) Beddoes was to find Oxford an inhospitable place,[3] and in 1793 he moved to Bristol to found his Pneumatic Institution.[4] Beddoes sought patients with respiratory disorders who might be cured by the use of various gases, and he knew that such sufferers flocked in their hundreds to the spas of Bath and Bristol (Hotwells). The latter place was probably chosen because of its greater radical sympathies, which were much needed when Beddoes' political campaigning against the Government, the rich and the Royal Society brought him national fame (or notoriety) and – the ultimate accolade for an avowed revolutionary – a hostile notice in *The Anti-Jacobin Review and Magazine*.[5] The occasion for this polemic outburst was not, as might have been expected, another political tract but rather a report on medical observations at the Pneumatic Institution.[6] Behind 'The Pneumatic Revellers. An Eclogue' lay a clear recognition of direct connections between Beddoes' scientific and political thought.[7]

Beddoes, as we have seen, was not the only man of science to espouse a fervent radicalism in politics. A majority of the Lunar Society and several of the influential members of the Manchester Lit. and Phil. are known to have held such views. None was more scientifically distinguished or better known than Priestley, and few paid a higher price. The Birmingham riots, whatever their immediate cause, were part of a larger movement of reaction, sparked off by fear that events across the Channel might be repeated in England.[8] When news of the worst excesses of the Terror became known the wave of natural revulsion which swept the country was augmented by a more synthetic indignation, orchestrated by those who might have most to lose. 'Jacobin' became a term of opprobrium, a convenient label which (for example) the Scottish land classes could attach to the Whig minority. The strongly polarised opinions of the 1790s were in turn replaced by less extreme attitudes, and at the same time revolutionary passion was often to mellow into reforming zeal. But political argument continued, and sedition was seen by Government as a real threat until the Reform Bill of 1832 and beyond. Science might seem to have little to do with such matters; it was, after all, making spectacular progress on its own front, with discoveries by Davy and the atomic theory of Dalton merely marking the high points of scientific progress. Yet, as *The Anti-Jacobin Review and Magazine* had perceived, a relationship may well exist between a man's science and his politics.

The purpose of this chapter is to enquire into the nature of that relationship in the case of the British radicals, and to ask four main questions. First, what personal links can be established between science and political radicals? Second, how, if at all, did radicalism affect the progress of science? Third, how, if at all, did anti-radical sentiment and legislation affect that progress? Fourth, can any ideological links be established between natural science and political radicalism?

In attempting to identify personal links between science and radicalism at this period nothing is easier than to provide anecdotal evidence for individual cases: names like Beddoes, Cooper, Henry and Priestley come readily to mind, and hosts of less famous individuals can be marshalled in support of the thesis that many practitioners of science had radical sympathies, or that many political radicals looked favourably on science. But for this fragmentary evidence to be persuasive it needs to be supplemented by quantitative evaluation, and this poses formidable problems. Defining the scientific community is difficult, though possible: one simply has to agree over criteria (e.g. taking an *active* part in founding or running a scientific enterprise, publishing papers, taking out patents etc.). Defining the radicals is much harder, partly because people's political tendencies change with time (today's radicals are often tomorrow's liberals), and, in the period we are discussing, were often kept secret for reasons of personal security.[9]

In Manchester the association between science and republican ideals has been noted already (p. 110). The prosopographic survey by Thackray, while giving quantitative data on religious affiliations of its members, does not unfortunately do the same for their politics. However its findings bear indirectly on the matter, as we shall see (p. 175). Detailed information for other provincial societies may be found in a paper by Ian Inkster dealing with the Mechanics' Institutes.

At Liverpool, according to Inkster, the leaders of the scientific community in 1820 included Thomas Bentley, James Currie, Arthur Heywood and John Yates, all of whom were 'committed Radicals prior to 1810' and 'at least verging on Republicanism in the 1790s'.[10] At Leeds, Edward Baines and William Wood were active in the local scientific community and strongly backed the Mechanics' Institute; they were Radicals in the 1790s and Whig supporters in the 1820s.[11] At Nottingham, the Philosophical Society of 1783 was supported by Charles Wilkinson who also promoted science at his Nottingham Academy and joined the radical Society for

Political Information in the 1790s.[12] All this information, though suggestive, lacks the force of quantitative data which Inkster does provide in the case of Derby and Sheffield.

The scientific movement in Derby was led by Edward Higginson, a Radical in the 1790s and activist in the Anti-Corn Law League, Reform Movement, and similar enterprises. Other reformers were Joseph and William Strutt, and William Evans, iron manufacturer. Of the thirteen men who attempted in 1812 to present the Petition for Peace to the Mayor, eleven were of the local scientific community and nine active in the foundation of the Mechanics' Institute. Taking the Derby and Sheffield Mechanics' Institutes together, and considering only the scientifically active members of each, from about 1825 to 1832, the political profile is as follows:

Radicals before 1820	23 per cent
Reformists 1820–1829	27 per cent
Reformists 1829–1832	57 per cent

It may be added that the 'scientifically active' reformists of the last period amount to 79 per cent of the known local parliamentary reform movement.

The situation in London was naturally different from the provinces, and also much more complex with all shades of political opinion to be found. 'Radical science' is most likely to be located in the records of fringe societies and institutions which London's large population was capable of sustaining. Various metropolitan institutions at about this time have been studied recently.[13] But much needs to be done before the extent of radical involvement becomes clear.

The London Chemical Society of 1824[14] is certainly typical in one respect, the paucity of its records. In fact the only detailed information available is contained in a periodical publication nearly as ephemeral as the society whose birth it chronicles. This was *The Chemist*[15] whose connection with the society was rather more than that of a mere reporter. On 26 May 1824 a letter appeared under the signature of 'A. W.' (later corrected to 'A. M.'),[16] in the following terms:[17]

> You must be aware, that the majority of the readers and admirers of your little Work is composed of persons who can

> devote but a small proportion of time to chemistry, and who at the same time are not in circumstances to lock up part of their capital in apparatus, &c. for prosecuting the study of that science; would it then not be advisable for you, who appear so much the encourager of beginners, to form, as it were, a nucleus for a society of young chemists, who might, at their common expense, purchase chemical tests, instruments, &c. as they want them, and that without bringing down ruin on any of them. I should like the society to be respectable and select, and to meet at a stipulated number of times every week.

The suggestion was taken up with alacrity in the correspondence columns of later issues, the anonymous editor expressed benevolent interest, a steering committee was established, and a Society duly constituted on 4 September 1824, with a half-yearly subscription of one guinea and an entrance fee of the same amount.[18] It was to have a lecture room and laboratory open daily on weekdays, and fortnightly meetings for lectures, demonstrations, 'the reading of memoirs' and so on. An official inauguration took place at the City of London Tavern on 12 November 1824, with at least three hundred persons present, to hear a declamation (one can use no other word) on chemistry and change, from the President, none other than George Birkbeck himself, physician to the Aldersgate Dispensary and pioneer in the Mechanics' Institute movement.[19]

Dr Birkbeck's oration bordered at times on the euphoric. The foundation of the London Chemical Society was 'under circumstances the most flattering and auspicious' and would certainly lead to great advantages. It is the more remarkable, therefore, that within three days of his lecture, a letter should be addressed to *The Chemist*, claiming disappointment with the Society because of 'lectures being delivered by persons who are not sufficiently pregnant with the matter to give information in an intelligible manner'.[20] And that is the very last reference to the Society which probably failed to survive even into the new year[21] and disappeared without trace, as did its literary counterpart a few months later.

The case for noticing such an ephemeral institution and

journal rests upon the radical interest in science which they revealed. In that sense their 'failure' is almost unimportant.[22] The connection between the two is made plain by the role of *The Chemist* in the Society's formation, by the evident relish with which its progress was reported, by the coy self-congratulation that 'perhaps had *The Chemist* never been published, the Society would never have existed'[23] and by the complete silence in which its untimely demise was passed over.

Information about personnel in the Chemical Society is limited to the report of the Inaugural Meeting. Nothing at all is known about membership, but Committee Officers are stated. They were:

The President:	Dr Birkbeck
Vice-Presidents:	J. F. Cooper, Esq., A. J. F. Marreco, Esq.
Treasurer:	G. Smith, Esq.
Secretary:	Mr W. Jones
Curator:	Mr J. B. Austin
Members of the Council:	Mr H. Fenner, Mr T. Dell, Mr W. S. Stratford, Mr H. J. Silva, Mr C. Dunderdale.

Birkbeck apart, most of these men are totally unknown in any scientific context. This is consistent with the view that the Society was a spontaneous grass-roots movement with scientific self-improvement as a genuine objective, as the initial correspondence suggests. Alternatively it could be construed as an expression of a radical political creed and as an instrument for social transformation. Key personnel might be known in political, but not scientific, circles in that case. Most probably, as we shall see, there was something of each.

Of the two Vice-Presidents J. F. Cooper defies positive identification. It is unlikely he was the J. F. Cooper, coal merchant, mentioned in the Directories.[24] More plausible is the suggestion that the name is a misprint for John Thomas Cooper (1790–1854),[25] a local chemical manufacturer and lecturer at Aldersgate Medical School, the Russell Institution, and the London Mechanics' Institute.[26] If so, he was the only chemist of any note on the Committee.

The other Vice-President can be definitely identified: Antonio Joaquin Freire Marreco. From 1821 to 1831 he appears as 'merchant' in the City, presumably selling wine from his native Portugal.[27] We then find him in the north of England until 1841 where he was Secretary of the Stanhope and Tyne Railway Company and later a Director of the Durham Junction Railway Company, two risky enterprises whose promoters included the Harrison[28] and Barnard[29] families.

The first of these ventures failed in 1841, one of 'the melancholy examples of unsuccessful management directed from Board-Rooms in London',[30] and Marreco had clearly had enough. In August of 1841 he resigned from the Board of the Durham Junction Railway[31] and is next heard of in 1842 as a Colliery Manager in Portugal.[32] There is no evidence that Marreco returned to Britain, though one son was born to him in Lisbon in 1849,[33] and his widow reappears in England in 1853.[34]

Marreco was a social climber and no radical. He married his employer's niece,[35] and, while retaining his mercantile activities,[36] he started a second career in railway administration, and invested substantial capital in railway development.[37] We do not know his age[38] but his ascription of 'esquire' suggests he had already made his mark by 1824. Possibly his business acumen, coupled with a genuine interest in science, made him an appropriate choice as Vice-President. Curiously, his son Algernon was later to achieve considerable distinction in the field of science education, as the first Professor of Chemistry at Newcastle College of Physical Science (later Newcastle University).

The Treasurer, G. Smith, is not identifiable, but he could have been a chemist and druggist of London Road,[39] or a chemist from Whitechapel Distillery who subsequently joined the 1841 Chemical Society,[40] or George Smith, MP (1765–1836), Banker of George Street, Director of the East India Company and a proprietor of the London Institution. Further certainty is impossible. The Secretary, W. Jones, sounds from his first letter in *The Chemist*[41] to have been an enterprising youth with a large room and a '*small* collection of books, utensils &c.' offered for general use. He does not appear in the

Directories at his stated address (55 Great Prescott Street) and so was probably a lodger. Of the other Committee Members, W. S. Stratford might, by an outside chance, be the astronomer of that name.[42] 'H. J. Silva', however, is almost certainly H. J. Da Silva, merchant, at the same address as Marreco from 1821 to 1832 and, from 1829 his business partner. The other three members are completely unknown. The Curator, J. B. Austin, was one of the lecturers praised by *The Chemist*[43] and fairly certainly John Baptist Austin, chemist and druggist of 3 Old Broad Street.[44]

So far the evidence for radical involvement is negligibly small; such positive identification as there is suggests a fairly youthful group enthusiastic for self-improvement, with a few slightly older men able to offer business or chemical experience. But there are still two missing factors. One is a name referred to only in the report of the Inaugural Meeting. This offers thanks to 'Mr Mongredieu, the Editor of *The Chemist*, Mr W. Jones the Secretary, the Committee, and the Chairman.[45] Assuming the comma is intended after the first name (and we know this person was not the Editor of *The Chemist*) we find a curious name that should be readily identifiable. The name spelt Mongredieu appears in local directories as Stationer, or Library, at 44 George Street, Hampstead Road.[46] With a slight variation of spelling, the name Mongrédien appears in papers of the radical organisation the National Political Union:

> 1831 Mongrédien, A, George Street, Newton Square: Council Member
>
> 1832 Mongrédien, A, merchant's clerk, 44 George Street, Hampstead Road: Ballot list for Council Election.[47]

This clearly unites him with the Mongredieu family of the Directories and therefore of *The Chemist*. He appears several other times in the Francis Place papers, especially as a member of the Radical Club on 21 January 1838. Here he is identified as 'Augustus Mongrédien, 21 Finsbury Square'[48] the writer, political economist and chess-player extraordinary.[49]

His father was a French Officer who fled to England on Napoleon's coup d'état in 1799 and was, presumably, in sympathy with French revolutionary ideals. Which Mongrédien is singled out for public thanks in The London Chemical Society does not really matter if we are merely seeking evidence for an injection of radical sympathies; but the probability must rest with the son. Who more likely than he could have been the 'A. M.' whose letter sparked off the whole affair? The tone of his first letter is that of a youthful enthusiast (Augustus was then 17), and a subsequent letter suggests he does not want direct involvement in running the Society,[50] which could explain both his omission from the Committee and the expression of gratitude by the inaugural meeting.

The second link we have not yet explored is that with *The Chemist*[51] itself. Of the radical views of that periodical there can be little doubt. Nowhere are these expressed more clearly than in the opening 'apology and preface', where the scientific establishment is roundly taken to task for its snobbery, exclusiveness and reluctance to promote scientific instruction amongst the working classes, to say nothing of its divisions into sects and parties. In no-one were these vices more evident, according to *The Chemist*, than in the President of the Royal Society, Sir Humphry Davy, whose portrait would not be allowed to adorn their periodical as a frontispiece:[52]

> As we have not observed any very great zeal, among those who are at the tip-top of science, to assist the working classes in the numerous and glorious efforts they have lately made to procure instruction for themselves, we confess a suspicion is excited that they look with no kindly eye on these efforts, and would rather have mankind for pupils than fellow-students of the great volume of nature. If this be correct, it might perhaps be an insult to our readers to place in the front of our pages the portrait of a man who, however learned, is not learned for their utility, and who seems to take little or no interest in their improvement.

'A constant reader' joined the attack on Davy with a scurrilous assault on his originality and even integrity. An editorial footnote sanctimoniously observed that complex apparatus was not necessary for good results in science, listing Franklin,

Watt and Priestley as shining examples of those who made valuable discoveries with 'a few pipe-bowls and common phials'.[53] Observing that 'the aristocracy of chemistry' would 'fain confine the science to their own class', the aggrieved correspondent yet rejoices that:

> the spirit of the age does not accord with the views of the dandy philosophers; they may black-ball at Somerset House, segregate from Albemarle-street, or shut themselves up in the atheneum [sic] . . . but they will only . . . have the mortification of seeing that the world goes on better without them.[54]

The publishers of *The Chemist*, Knight and Lacy, were publishers also of the *Mechanics' Magazine* and other literature for the working classes.

A similarity between the *Mechanics' Magazine* and *The Chemist* is not surprising in view of the identity of the always anonymous editor of the latter. This was revealed by Henry Brougham in his *Practical Observations upon the Education of the People*, where, praising *The Chemist* for its 'admirable collection of the most useful chemical papers and intelligence', he announced that it was 'learnedly and judiciously conducted by Mr Hodgkin'. Despite a mis-spelling Brougham can only have meant one man in that reference, Thomas Hodgskin, radical economist, political writer, anticipator of Karl Marx, and an activist in the early Mechanics' Institute Movement. In fact, Hodgskin had been a co-editor of the *Mechanics' Magazine* when it first appeared.[56] A toast at its Annual Dinner on 2 December 1824 included both Robertson and Hodgskin as 'the enlightened editors', but a somewhat embarrassed editorial note by the surviving editor indicated that 'Mr Hodgskin has for a considerable time ceased to have any connection with the *Mechanics' Magazine*'.[57] We can see why: he had moved across to manage *The Chemist*. He was frequently at odds with his colleagues, for example, Place reckons both he and Robertson 'knew next to nothing of the habits and feelings of London workmen' and deplored his intention to lecture The Mechanics' Institute on his 'hobby' of political economics.[58] By the end of 1824 Hodgskin was displaying special interest in chemistry, writing of 'the mechanic in the gas works' who was 'a chemist of considerable skill, and converted a quantity of coal, by his chemical skill, into products of ten times their value.'[59]

In Thomas Hodgskin and Augustus Mongrédien we have positive evidence of the radical underpinning of one small metropolitan scientific enterprise. But in Marreco, Silva, Austin and others we have no less clear an indication of a mercantile and technical competence with which the former group may have had little in common. Ultimately it was not a question of comprehension of specific scientific facts so much as widely different value judgements as to the way in which science should be used. Perhaps that is why both *The Chemist* and The Chemical Society ultimately failed. A common devotion to youthful improvement in chemistry was simply not enough to hold different ideological factions together. It is hard to see from this episode how the knowledge and promotion of science was significantly advanced by radicalism. But it shows they were probably in association.

Other cases exist where science was demonstrated as part of radical culture in London.[60] In medicine there were James Watson, the apothecary John Gale Jones, and the Hoxton physician who established an international reputation in palaeontology and gave his name to a nervous disease, James Parkinson.[61] The pioneer of speech therapy and ubiquitous lecturer, John Thelwall,[62] was considered sufficient of a menace to be arrested on a charge of high treason in 1794, though he was subsequently acquitted.[63]

An alliance between science and radical politics when radicalism was unpopular and even feared may have caused science actually to suffer from the association. There are several instances where this was so, none more obvious than the Parliamentary proscription of scientific meetings by a succession of repressive Acts between 1795 and 1819.[64] These Acts prohibited meetings of certain explicitly political kinds, and regulated assemblies of non-chartered societies by a system of licensing. Records of the licences, and of prosecutions for illegal assemblies, constitute an important historical source that has only recently been explored.

The fears that promoted such legislation in the 1790s were reactions to the French Revolution; the Seditious Meetings Act of 1817 was a response to the post-war surge of revolutionary sentiment and action;[65] its five-year renewal in December 1819 followed the tragedy of Peterloo.

In the event, the effects on science of the restrictive legislation do not seem to have been very great. Inkster has pointed out the strength of parliamentary opposition before the Bill became law,[66] and Weindling discovered that (in Middlesex) only a small fraction of science lecturers were either licensed or prosecuted.[67] In the first instance the government used institutional rather than epistemological criteria, asking whether the societies were chartered rather than whether they were 'scientific'. There were considerable difficulties in deciding whether a society was 'philosophical' or 'political.'[68] Inkster's belief that the exclusion of religion and politics from the official agendas was a matter of survival[69] may conceal the very diverse nature of such societies.[70] It is also noteworthy that such exclusiveness prevailed in provincial societies as the Manchester Lit. and Phil. before restrictive legislation was passed.

It should not be imagined that radical involvement with science was chiefly regulated from outside. Government legislation was much less effective in practice than the efforts of the scientific community itself. Most notably, successive Presidents of The Royal Society, in seeking to maintain its monopolistic hold on British science, opposed all efforts to set up 'rival' organisations, and that of course included those with radical sympathies. Under the Presidency of Sir Joseph Banks this monolithic structure was strenuously maintained. (The only concession Banks would make to the increasing demands for specialisation was to permit 'Assistant Societies' such as 'The Animal Chemistry Club',[71] which would only operate under his control.) Admission as Fellow of the Royal Society was denied to several known radicals, notably Thomas Cooper of Manchester.[72] The case of Thomas Beddoes is instructive. Banks steadfastly refused any kind of support for the Pneumatic Institution, and most of the London scientific establishments followed his example. This can easily be interpreted as hostility to Beddoes' radical views. 'I suppose he has seen Beddoes' cloven Jacobin foot', was how the younger James Watt saw it. But in fact Banks claimed to be 'wholly satisfied' with assurances that Beddoes was not a public menace, and said that his doubts now centred upon the probable dangers to the patients being subjected to Beddoes' experiments.[73]

When Banks died in 1820 he was succeeded as President of the Royal Society by Sir Humphry Davy. Long ago Davy had enjoyed the company of radicals, who, Bristol fashion, had crowded the house of his mentor Thomas Beddoes. If ever he had been touched by their republicanism, his writings do not show it, though they certainly reflect the prevailing romantic attitude to nature. A few years later, in 1802, his political *credo* was expressed in this optimistic passage from a 'Discourse on chemistry':[74]

> The arts and sciences also are in a high degree cultivated, and patronized by the rich and privileged orders. The guardians of civilization and of refinement, the most powerful and respected part of society, are daily growing more attentive to the realities of life; and, giving up many of their unnecessary enjoyments in consequence of the desire to be useful, are becoming the friends and protectors of the labouring part of the community. The unequal division of property and of labour, the difference of rank and condition amongst mankind, are the sources of power in civilized life, its moving causes, and even its very soul.

By 1820 Davy was inclined to Whig views and launched his Presidency from a platform of reform, institutional rather than political, but reform none the less. Fellows were to be chosen from a wider section of society than before; scientific expertise was to be a more stringent condition and the concept of appropriate 'science' was to be broadened to include, for example, geology and mathematics; other societies were to be encouraged, as were individual workers; and the Royal Society was to initiate stronger links between the government and the scientific community.[75] But by no stretch of the imagination could Davy's measures be regarded as radical, and it is not hard to see why he was so lampooned in *The Chemist*. From the perspective of that journal, reform was as irrelevant as reaction.

Davy's later years were plagued by ill-health, and in 1827 he resigned his Presidency. His place was taken by a former Cornish associate Davies Gilbert (formerly Davies Giddy) who, as MP for Bodmin since 1806, had looked after the Royal Society's interest in Parliament. Although once a supporter of radical politics himself and a friend of Beddoes (to whom he had introduced Davy), Gilbert was now a government supporter, and he continued Davy's policies for the Royal Society. His Presidency lasted from 1827 to 1830.[76]

Radical science was also attacked by that other wing of the intellectual establishment, the Universities. The expulsion of Beddoes from Oxford has already been noticed (p. 136), and individual academics of considerable influence used their voices and pens to discredit what they regarded as dangerous tendencies in science. Perhaps the most famous case was the Professor of Natural Philosophy at Edinburgh, John Robison (1739–1805) whose patron and colleague had been Joseph Black in Glasgow. Robison himself was Secretary to the Royal Society of Edinburgh and a respected man of science. In 1797 he published a book that was to bring him additional fame, *Proofs of a Conspiracy Against all the Religions and Governments of Europe, carried on in the secret meetings of Free Masons, Illuminati and Reading Societies*; it went into four editions in just over a year and was widely circulated in Britain and abroad. It attempted to show that in France and Germany Masonic Lodges had been taken over by radicals, 'Illuminati', and that in Britain political and religious stability was similarly under threat. Singled out for particular demolition were the ideas of Joseph Priestley, partly on account of his political radicalism, partly because he tried to explain mental phenomena in terms of a materialist ether, and partly because he seemed to reduce the deity to undulation in that same ether. All this was made a great deal worse by Priestley's impious 'borrowing' of the ether from the devout Newton. Here indeed was an interesting case in which politics, religion and science seemed almost inextricably interwoven, and 'Newtonianism', at least in some senses, was once more a polemical tool for an ideology that embraced much more than mere science. In fact some of the precautionary measures urged by Robison such as censorship, banning secret societies etc., were introduced piecemeal by

Pitt during the 1790s, and evidence exists that notice was taken of his book by the Lord Advocate of Scotland.[77]

In 1799 Joseph Black had died and four years later Robison produced an edition of Black's Lectures. Finding that the rather scanty notes left by his patron suggested a certain favourable view of Lavoisier's chemistry, Robison carefully edited them so as to highlight Black's opposition to other aspects of the chemistry which he (Robison) quite wrongly associated with Jacobinism. A later attack on French science focussed on the atheism which he thought was implicit in the writings of Laplace. Nor was Robison alone in such attitudes. In contributing to the Supplement to the third edition of the *Encyclopaedia Britannica* he co-operated with the senior science editor, George Gleig, in producing numerous articles in which no opportunity was lost to attack ideas of materialism or determinism which they regarded as characteristically French, probably Jacobin and certainly obnoxious.

The polemics of Robison and his friends are enough to caution us against too facile an identification of the interests of science and political radicalism. Certainly there are superficial similarities in their approach: rejection of imposed authority, rationality, recognition of potential for social change, openness of data, and so on. Where these are also stressed by a religious creed, like Unitarianism or Quakerism, their mutual reinforcing effect is always stronger. But none of these morphological similarities is complete, and so the paradoxical situation arises in which science can be pursued by both the establishment and its enemies, and can be used both to support and to attack radicalism.

In 1791 Priestley wrote: 'the English hierachy (if there be anything unsound in its constitution) has . . . reason to tremble even at an air-pump or an electrical machine'.[78] But it was not to be. These very instruments of science were in fact to be taken up and used by the political establishment, at least at local level, for their own ends. The manner in which this happened will be the subject of the next chapter.

1. Priestley's chemical apparatus. Showing the various devices used to prepare and collect gases.

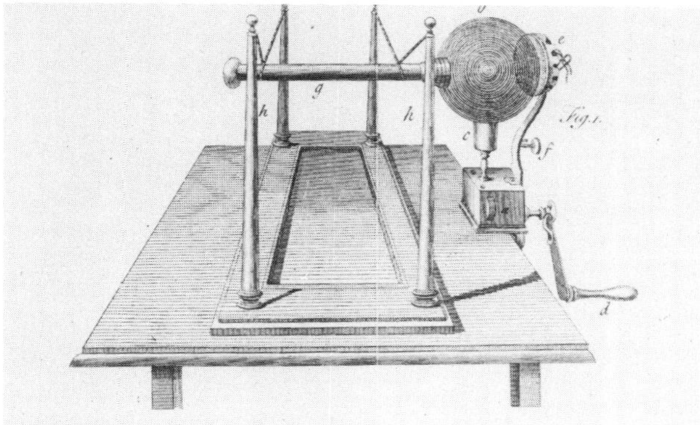

2. Typical eighteenth-century electrical machines. Static electricity was generated by rotating the glass jar or globe against a 'rubber'.

3. An idealised representation of the study of natural history in the early eighteenth century (Levin Vincent, 1790).

4. 'The starry heavens': another idealised picture, showing telescopic observations of the heavens. Note the symbolism of piety and Newtonianism.

5. Instrument maker's workshop in the early eighteenth century.

6. Chemical laboratory in early eighteenth-century London, thought to be that of A. G. Hanckinitz, a wealthy amateur and former assistant of Robert Boyle. Note the air pumps, characteristic of mechanical philosophy, the retorts inherited from alchemy and the balance and furnaces of the assayers tradition.

7. Manufacture of sulphuric acid, an eighteenth-century engraving. Note the large, fragile glass globes in which the acids reacted.

8. One of many cartoons directed at Priestley. Note the combination of chemistry, politics and theology.

9. The sacking of Priestley's house by a mob during the 'Church and King' riots of 1791.

10. Gilray cartoon of the early days of the Royal Institution. The central figure between the rostrum is Dr Garnett, assisted on the right by Humphry Davy. The entertainment appears to consist of administration of laughing gas (nitrous oxide). Air bladders, bell jars and other pneumatic apparatus may be seen through the doorway and an electrical machine. The audience in 1801 does not yet seem to have been the fashionable one of later years.

11. Title page of the *Mechanics' Magazine* (see p. 155).

12. The Manchester Mechanics' Institution, erected 1825.

13. Laboratory of London Mechanics' Institution, 1828. Although it appears to have gas lighting, the most notable feature is its smallness and the simplicity of its equipment. Judging by its array of bottles, chemistry seems to have been the chief science studied here.

14. The Sheffield Athenaeum and Mechanics' Institute.

15. The Leeds Mechanics' Institute, 1865.

16. Exterior of the Newcastle Literary and Philosophical Society.

17. Royal College of Chemistry, London, 1846.

18. The Felling Chemical Works, Newcastle. The print shows the soda furnaces.

19. Leblanc rotary furnace, made and extensively used on Tyneside from the 1860s onwards.

20. The Great Exhibition, 1851: southern entrance to the transept.

21. The Royal Commissioners for the Exhibition of 1851: from left to right standing are Dilke, Scott Russell, Cole, Fox, Paxton, Lord John Russell, Peel, Robert Stephenson; left to right seated: Cobden, Barry, Lord Granville, Cubbitt, Prince Consort, Lord Derby.

22. Darwin's funeral, 1882: this sketch in Westminster Abbey depicts many leading men of church, science and state.

9. Science for the Masses, 1825–1850

9.1 THE DEMOLISHED STAIRCASE

THE design of the now famous Lecture Theatre at the Royal Institution, London, ought to have been an uncontroversial affair.[1] By 1801 the building was completed, very like the theatre in the Andersonian Institution, Glasgow, from whence had lately come the Royal Institution's first professor, Thomas Garnett. It was, however, the brain-child of the Clerk of Works, Thomas Webster (or so he claimed). Aware of the spirit of philanthropy in which the Royal Institution had been conceived, Webster was anxious to establish a school for mechanics, enabling them to understand the rudiments of science and mechanical drawing. As part of that scheme, which was coolly received by the Managers, he included in the theatre a gallery, 'intended for those who either wished to be less observed, or who, for obvious reasons, would not like to sit down by their employers'.[2] This gallery was reached by a separate stone staircase which led directly from Albemarle Street without communicating with any other part of the building; in this way the mechanics could come and go without meeting anyone else. This is where the controversy began. Webster wrote:[3]

I was asked rudely (by an individual whom I shall not now

> name) what I meant by instructing the lower classes in Science? I was told likewise it was resolved upon that this plan must be dropped as quickly as possible. It was thought to have a dangerous political tendency. I was thus told if I persisted I would become a marked man!

Shortly afterwards the management of the Institution changed and, as Webster sadly recalls, 'from what appeared to me very erroneous reasoning my mechanics' stone staircase was pulled down at a considerable expense'.[4]

The demolition of the staircase symbolised, though it did not cause, the rapid elimination of artisan training from the programme of the Royal Institution. It was also symptomatic of a widespread English antipathy to popular scientific education that, for much of the century, barred most of the population from access to scientific knowledge. Fortunately, however, other routes were available, and even if the footholds were less secure not a few of the population were able to arrive at a first-hand knowledge of science.

Apart from the Dissenters' Academies, no English or Welsh schools of any kind made provision for scientific training until well into the nineteenth century. When it did happen, one of the earliest pioneers was Richard Dawes (1793–1867), Cambridge graduate, Fellow of Downing College, and a man of liberal views, supporting the cause of the nonconformists.[5]

Appointed in the late 1830s as Rector of the country parish of King's Somborne, Hampshire, he was immediately faced with an urgent need for educating his parishioners. When National Society (i.e. Anglican) schools were established there, he devised curricula connected with everyday occurrences, and began teaching the 'science of common life', continuing and greatly extending an approach that a few others had tried before him (notably Charles and Elizabeth Mayo at Cheam School). His teaching of applied science coincided with a growing national awareness of the relevance of science to agriculture, and was further reinforced by the proximity of another remarkable educational establishment, Queenwood College, about four miles to the west.[6] This school, founded in 1847 by the Lancashire Quaker George Edmondson, became one of the first in England to teach experimental science. Links between King's Somborne and Queenwood existed in the person of Edward Frankland, later to pioneer popular science education on a much larger scale. In 1850 Dawes became Dean of Hereford, but his work at King's Somborne was repeatedly praised for pointing the way for future developments in school science education.

In the first half of the century the vast majority of English school children had no chance whatever of encountering science at school. For a diligent and fortunate few some opportunity did exist in popular or semi-popular textbooks that might be found in the subscription libraries or at home.[7] Most famous, perhaps, was Mrs Jane Marcet's *Conversations on Chemistry*, running to no less than eighteen editions from 1806 and selling 160,000 copies in the United States alone. She was the wife of A. J. G. Marcet, a chemical lecturer at Guy's Hospital London,[8] and initially wrote for the fashionable audiences at the RI. Nevertheless this was a book that won over to chemistry a young bookbinder's apprentice, Michael Faraday.[9] In similar conversational form were the *Scientific Dialogues* of the Reverend Jeremiah Joyce which, from 1809, made a deep impression on many working lads,[10] while the young Charles Darwin was one of the many boys who encountered chemistry by carrying out the experiments described in the *Chemical Catechism* of Samuel Parkes (Twelfth Edition from 1806).[11] More surprisingly, that work of excruciating sentimentality but great popularity by Thomas Day, *The History of Sandford and Merton* (1783 on) was the introduction to science for both Edward Frankland[12] and A. R. Wallace.[13] Others later to climb from humble origins to the pinnacles of British science were not so fortunate. T. H. Huxley at the age of twelve was confronted with Hutton's *Geology*[14] while Humphry Davy had to struggle with Lavoisier's *Traité Élémentaire de Chimie* in French, with Nicholson's *Dictionary of Chemistry* (1795) as a reference standby.[15] The Lancashire pioneer of textile chemistry, John Mercer, taught himself from a secondhand copy of James Parkinson's *Chemical Pocketbook*, picked up in Blackburn market.[16] Joseph Swan, inventor of the incandescent electric light, came across 'a little book of rudimentary chemistry' by Hugo Reid.[17] Charles Babbage arose at three a.m. to read Ward's *Young*

Mathematician's Guide.[18] Rather later, Ambrose Fleming, inventor of the thermionic valve, was captivated by J. A. Pepper's *The Boy's Playbook of Science* (1860), a work more obviously written for a juvenile readership.[19]

Books might be encountered by chance, but many of those who achieved success had access to libraries. Davy may well have used the Book Club in Penzance, or the library of his relative John Tonkin.[20] Herbert Spencer had access to the Derby Philosophical Society Library[21] (his father was the Honorary Secretary) and the local Methodist Library. Frankland, like many others, was able to use the libraries of the new Mechanics' Institutes (of which more later), particularly recalling in later life Priestley's *History of Electricity* as well as Joyce's *Scientific Dialogues.*[22]

Given the availability of books of this kind it was certainly possible for unprivileged youths to acquire information about science, and even the stimulus to pursue it for most of their working lives. But one cannot argue from the successful few that such facilities were widely used. Had these been the only sources of systematised natural knowledge available to the artisan, his prospects would have been bleak indeed. In the 1820s, however, two related movements appeared that held promise of better things: The Mechanics' Institute Movement and the Society for the Diffusion of Useful Knowledge. In the event neither turned out to be quite the grand stairway to knowledge that had been hoped at first. A temporary ladder might be a more appropriate metaphor. In both the rise and decline of these movements scientific knowledge was closely linked with intended and actual changes in English society.

9.2 NEW ROUTES TO SCIENCE IN ENGLAND

The benevolent intentions for artisan education at The Royal Institution came, as we have seen, to nothing. Then, in 1823/4, were felt the first stirrings of a movement to transform the shape of British society. It did just that, though not quite in the way its founders intended.

In 1800 the Edinburgh-trained physician George Birkbeck

had conceived the brilliantly successful plan of providing science lectures for mechanics. He wrote:

> Whilst discharging the duties of Professor of Natural Philosophy and Chemistry, in Anderson's Institution, in Glasgow, I had frequent opportunities of observing the intelligent curiosity of the 'unwashed artificers', to whose mechanical skill I was often obliged to have recourse; and on one occasion, in particular, my attention was arrested by the inquisitive countenances of a circle of operatives, who had crowded round a somewhat curious piece of mechanism which had been constructed for me in their workshop. I beheld, through every disadvantage of circumstance and appearance, such strong indications of the existence of the unquenchable spirit, and such emanations from 'the heaven lighted lamp in man', that the question was forced upon me, why are these minds left without the means of obtaining that knowledge which they so ardently desire, and why are the avenues to science barred against them, because they are poor?[23]

On successive Saturday nights the numbers attending were 75, 200, 300, and 500, and for the next three years enrolments were around the 300 mark.[24] When Birkbeck left Glasgow for a more remunerative occupation as a physician, the class continued under Andrew Ure until 1823, when, in a bid to gain more democratic control by its members, it seceded from the Andersonian, and became the Glasgow Mechanics' Institution.[25] Meanwhile in 1821 a School of Arts had been established in Edinburgh with very similar aims,[26] and libraries for mechanics soon appeared in Liverpool, Sheffield, Kilmarnock, Greenock and elsewhere.[27] The *Mechanics' Magazine* first appeared in 1823, and it was this that was to catalyse the explosion that followed. The editors, Robertson and Hodgskin, proposed a metropolitan analogue to the Institution in Glasgow, 'the London Mechanics' Institution'. Birkbeck, now living in London, seized on the opportunity, and under his Chairmanship a steering committee was formed as a result of which, early in 1824, the new Institution opened its doors to those mechanics able and willing to pay the one guinea annual subscription. Within two years over one hundred similar institutions had appeared, ten in the London area, about forty in the industrial north, and nearly thirty in Scotland.

Hard on the heels of the Mechanics' Institute Movement came the Society for the Diffusion of Useful Knowledge,[28] called into existence by Henry Brougham, a friend of Birkbeck, politician, and former student of natural philosophy at Edinburgh. In 1825 he published his *Practical Observations upon the Education of the People, addressed to the working classes and their employers.* It was dedicated to Birkbeck and proposed to reinforce his efforts by creating a society which could sponsor and distribute cheap and useful books. In the next few years his Society produced a series of specially written textbooks under the general imprint of the *Library of Useful Knowledge*, and also the famous *Penny Cyclopedia.* The SDUK was not technically a publisher, but a patron to whom authors were contracted.[29] Like Birkbeck, Brougham relied to some degree on charitable support, and from then on the two movements complemented each other.

The later history of the Mechanics' Institute Movement and of the SDUK has been frequently discussed.[30] Individual institutions and local groups have been extensively studied, though much remains to be done.[31]

The one thing on which all writers agree is that, by and large, the Mechanics' Institutes were a colossal failure in the sense that the constituency commonly assigned to them remained largely untouched once the initial flush of enthusiasm has faded. Moreover as seedbeds of technological and scientific research they were conspicuously unsuccessful. There, however, unanimity is at an end, and differences of opinion exist about the reasons for the decline in working class support, about the extent of it and about the value of the alternative kind of institution that emerged.

First, the deterrent effect of subscriptions undoubtedly contributed to that decline, especially in times of high unemployment and falling wages (as in the depression of 1826).[32] Secondly, it is probable that lectures were rarely at an appropriate level for an audience of artisans. One has only to read a sentence or two from a peroration by Birkbeck to wonder what on earth the mechanics could make of it. The tendency to replace science by more elementary instruction suggests how great was the intellectual gap between what was expected and what was possible. Although literacy figures are notoriously unreliable, it appears that in 1840 about 75 per cent of the adult population had some reading ability and about 80 per cent some writing ability, though less than half were competent in arithmetic.[33] But it certainly did not follow that lecturers were blameless. As funds ran low, the most professional would be priced out of the market. With unpaid volunteers the prospect was even more problematic. Not all would have the experience of Rugby Mechanics' Institute of history lectures from Dr Thomas Arnold, with two great coloured charts as visual aids.[34] Thirdly the sheer exhaustion of the twelve-hour working day must have deterred all but the most strongly motivated from an evening lecture in an ill-ventilated room.

More controversial are those explanations which depend upon cultural values. It was suggested by Engels that the Mechanics' Institutes were taken over by the bourgeoisie, so that 'the mass of working men naturally have nothing to do with these institutes, and take themselves to the proletarian reading rooms and to the discussion of matters which directly concern their own interests'.[35] This does not explain why the takeover occurred and the workers were alienated. Moreover it ignores the fact that in some cases (e.g. Sheffield) the Institutes were 'middle-class from the very beginning'.[36] However, it is certainly true that religious and political controversy were excluded, at a time when the political interests among the working-class were mounting. The development of working class consciousness would certainly have been inhibited; one mechanic complained: 'you must remember we have masters all day long, and we don't want 'em at night'.[37] In some cases there was suspicion that 'the masters had some secret motive'.[38] The hard economic fact was that without financial support from the 'masters' the Institutions had no prospect of survival, and so the paternalism of a Birkbeck was continued as the patronage of the employers. To those seeking cultural as well as some measure of economic independence the prospects were simply not attractive enough.

It was once customary to picture the decline in working class support for the Mechanics' Institutes as uniform and nearly complete. Recent work emphasises the ambiguity of some evidence previously cited in support of the traditional view.[39] Contemporaries of the early Mechanics' Institutes by no means spoke with one voice. As Royle convincingly argues, 'the occupation of a mechanic does not, in itself, indicate that a man felt himself to be a member of a distinct social class, with the thoughts and prejudices of a proletarian; neither does the occupation of clerk or warehouseman suggest that such individuals had more in common with the masters than with the operatives'.[40]

The movement was never quite as monolithic as some had made it appear to be. The great variety of institutions is now beginning to be appreciated, and it is very apparent that the experience of the London Mechanics' Institution was by no means typical of the provinces, and that wide variations existed there also.[41]

By 1849, only 43 of the 204 Mechanics' Institutes in England and Wales were supported mainly by artisans – just over 20 per cent, it is true, but still a significant minority. Worker control must be distinguished from worker participation, but even here the picture is complex. In London worker control was initially intended but a compromise scheme emerged in practice. In Yorkshire the workers were in control at Bradford and Sheffield but not at Leeds.[42] In Manchester, as in Glasgow, a breakaway institution was founded on this very principle.[43]

Yet when all this is said, there remains the most fundamental ambiguity of all – what is, and was, meant by a 'Mechanic'? According to the London Mechanics' Institution he was 'every man who earned his living by the work of his hands',[44] so no sense of incongruity would arise amongst Birkbeck's companions on the Committee: four engineers, two smiths, one carpenter, one oilman, one printer, one coachmaker, one painter and glazier.[45] In so far as this was the kind of membership generally intended in the early days of the movement, one sees immediately that the 'decline' can be greatly exaggerated. These men were not the unskilled machine minders in the textile factories, but the skilled or semi-skilled manual workers. They were the 'aristocracy of labour', the people who were often in the first place the intended clientele.[46]

The effectiveness of the Mechanics' Institutes can be measured in terms of the stated objectives for 'mechanics' (even in the wider sense of the term), in which case it must be reckoned to be fairly low. When it is recalled that science and the useful arts were intended as a chief object of instruction, the sense of failure is deepened still further. Because of the inherent difficulties in teaching science under the prevailing conditions, sustained systematic teaching became replaced by short courses. It was not that science had been tried and found wanting, but rather that it had been tried and found difficult. Hence there was a shift toward lighter science (even scientific entertainment) and away from science altogether. It can be seen by an analysis of book holdings of the Institutions where the proportion of science books (though not necessarily their number) declined almost everywhere.[47] In 1843 10 per cent of the membership of the Yorkshire Union of Mechanics' Institutes attended classes in the 'three Rs'; in 1859 their elementary classes far outstripped all others.[48] Yet, as Royle observes, a failure to teach science 'in no way invalidated the work being done in the elementary classes'. To have provided a second chance for elementary education for thousands of working people of all kinds was one of the redeeming features of the movement. Set against the widespread failure of the schools to teach literacy and numeracy, these achievements cannot be discounted merely because they were not originally intended.[49]

By 1850 there were in England 700 institutions with a membership of 107,000 and a total holding of 690,000 books.[50] The annual circulation figures were over 1.8 million with a total population in England and Wales in 1851 of just under 18 million; this represents one book per ten members of the population.[51] With the advantage of hindsight one can also appreciate the continuity of educational opportunity afforded by these institutions, some of which became, in due course, prestigious centres of advanced education in our own day. Some have developed into modern polytechnics (Huddersfield, Leeds, Liverpool, Preston), and others into University institutions like Birkbeck College (London) and the University of Manchester Institute of Science and Technology. In their own day they became the nurseries for many who were later to distinguish themselves in science. Indeed, it has been argued that the Mechanics' Institutes were, in one respect, 'too successful', diverting a considerable number of working men 'into the paths of social ambition'.[52] To take one single example, the small and unimpressive Mechanics' Institute of Lancaster played a crucial role in the early development of Sir William Turner (eventually Vice-Chancellor of the University of Edinburgh), Sir Ambrose Fleming, Sir Edward Frankland (one of the most important British chemists of the last century), George Maule (the manufacturer of aniline dyes) and several others who went on to achieve academic eminence in the scientific world. The point here is not that the Lancaster Mechanics' Institute was uniquely responsible for the intellectual growth of these people, but that it was part of a larger local network in which young people in the town were encouraged to pursue scientific studies.[53]

9.3 THE RHETORIC OF USEFUL SCIENCE

We must now enquire into the rationale behind the Mechanics' Institute and SDUK movements. At first this appears to be a perfectly straight-forward expression of Benthamite Utilitarianism. It was clearly stated in the Annual Report for 1828 of the Manchester Mechanics' Institution:[54]

> The Manchester Mechanics' Institution is formed for the purpose of enabling mechanics and artisans, of whatever trade they may be, to become acquainted with such branches of science as are of practical application in the exercise of that trade. . . . It is not intended to teach the trade of the machine maker, the dyer, the carpenter, the mason, or any other particular business.

Similar aspirations were enshrined in documents for dozens of other institutions. There was, of course, a generally held belief amongst social reformers that education could 'break the bondage of poverty' by enabling people to become 'upright and useful members of the community, . . . illustrious examples of gratitude and virtue'.[55] This optimistic philosophy, coupled with a touching faith in the 'improving' efficacy of science, gave to those who believed in it an irresistible urge to treat the populace to as strong a dose of science education as it would take.

Exactly how a course of lectures on hydrostatics or the steam-engine should be 'useful' to mechanics is not altogether clear. Birkbeck seemed to think the answer lay in some kind of personal satisfaction:[56]

> I have become convinced that much pleasure would be communicated to the mechanic in the exercise of his art, and that the mental vacancy which follows a cessation from bodily toil, would often be agreeably occupied by a few systematic philosophical ideas, upon which, at his leisure, he might meditate.

How far such intellectual delights would appeal to the average mechanic must remain problematical. As Cobbett remarked 'the expansion of the mind is very well; but, really, the thing which presses most, at this time, is the getting of something to expand the body a little more: a little more bread, bacon, and beer'.[57]

Henry Brougham considered that science would enable a workman to do his job more efficiently:[58]

> in truth, though a man be neither mechanic nor peasant, but only one having a pot to boil, he is sure to learn from science lessons which will enable him to cook his morsel better, save his fuel, and both vary his dish and improve it. . . . But another use of such knowledge to handicraftsmen and common labourers is equally obvious: it gives every man a chance, according to his natural talents, of becoming an improver of the art he works at, and even a discoverer in the sciences connected with it.

In suggesting that scientific knowledge itself might be advanced, Brougham went a good deal further than Birkbeck, who was by no means sure 'that by a course of instruction . . . one artist will be directed to the discovery of anything which is essential or important in his particular department'.[59]

Whether in response to this kind of rhetoric or not, the mechanics undoubtedly saw in education a means of self-advancement, and this was especially true in science education. Utilitarian beliefs in science were not the prerogative of a few wealthy philanthropists.[60] The first cover of the *Mechanics' Magazine* was decorated with the names of ten pioneers in technology, most of whom were Fellows of the Royal Society and all but two (the Earl of Stanhope and the Duke of Worcester) of fairly humble origins though Watt and Smeaton were from the middle classes. Engineering predominated (Watt, Stanhope, Smeaton, Fulton and Rennie) though Rumford, Priestley, Newton and the instrument-maker Ramsden were also included. Science was represented as having a very practical outcome. The Baconian caption 'knowledge is power' undoubtedly caught the imagination of the readers who could in some measure identify themselves with the heroes paraded before them.

Several writers have argued that the projectors of the Mechanics' Institutes had more in mind than considerations of utility. Thus Shapin and Barnes suggest 'the initial stress of the Institutes on 'pure' science and 'scientific principles', their general neglect of applied subjects and practically relevant knowledge, and their total failure to develop (or even seriously to plan for) actual technical research, does cast doubt on the extent and immediacy of genuine utilitarian concerns as motivating their foundation. Moreover, the claim that teaching 'principles' made more creative and innovative employees comes oddly from those who, at the time, were obsessed with inculcating in their labour-force a rigid conformity and docile acceptance of routine.'[61]

There are really three arguments here. First an incompatibility is implied between a rhetoric of utility and the study of 'pure' science. This presupposes that we can read into the early nineteenth century our modern distinction between pure and applied science, whereas the distinction made by Birkbeck and his colleagues, quite openly, was that between teaching science and teaching a trade. The model of science they had was not two semi-autonomous areas of pure and applied science, but rather a body of generalised knowledge (science) which could then be used to inform, illuminate and even inspire the execution of practical projects. There was therefore a clear logical purpose in teaching science as then understood, and this is accurately reflected in both the lecture courses originally conceived and in the early library lists. Science could not be applied until it was understood. And it could only be understood by putting in a great deal of very hard work. This may be an uncomfortable attitude today, as well as an unpopular one then, but it was not necessarily inconsistent with the strategy adopted.

Secondly, we are reminded of the absence of actual technical research. Here again there may be a suspicion of anachronism. Research, to us, does need to be planned for, organised and perhaps guided in a project. But, as the list of 'heroes' suggests, research in the early nineteenth century was an intensely individual thing, to be conducted in a spirit of free enquiry. If occasional research was undertaken in this way we should not expect the Mechanics' Institutes to sponsor it nor their records to show it. More important is the fact (or the belief) that knowledge must precede explanation. Had systematic instruction continued it is possible that research might eventually have developed, but the fore-shortening of the whole teaching programme precluded anything of this kind.[62] The possibilities of research were further diminished by the sheer cost of laboratory facilities. One of the few places in Britain where teaching and research laboratories were provided at about this time was Thomas Thomson's Department of Chemistry at the University of Glasgow. In 1831 that cost the University £5000.[63] Certainly the larger Mechanics' Institutes (as at Manchester and London) had their own laboratories, but they were small,[64] and their use is largely unknown. In Lancaster a small laboratory was made available outside the Institute by a local medical man, Christopher Johnson, but this kind of provision may have been rare.[65] At a time when few works laboratories existed, it is hardly surprising that the promoters of Mechanics' Institutes did not appear to encourage original research.

But even if laboratories had been available apparatus was

expensive and often unsuitable. In 1824 the large sum of 250 guineas was spent on apparatus for the London Mechanics' Institution,[66] even more than John Dalton's £200 order of research apparatus a few years earlier.[67] Yet much of this must have been unsuitable for serious work for the relatively unskilled, as witness Birkbeck's grateful letter to the Glasgow apparatus manufacturer J. J. Griffin, who was profusely thanked for the apparatus which could now be used for an extensive series of class experiments, thus providing the best means for extending the theory and practice of chemistry. But that was as late as 1838.[68]

In the third place it has been suggested that enlightened education of the mechanics was incompatible with an obsession by their employers 'with inculcating a rigid conformity and docile acceptance of routine'. Apart from a certain circularity of argument (since the main purpose of Shapin and Barnes seems to be to establish that the employers were in fact doing that), this point ignores the large number of mechanics and their radical friends who *were* using the rhetoric of utility. They were presumably not entertaining the same obsessions. Now if the employers *could* see gain as well as loss for themselves in the advancement of their work-force, their rhetoric of utility might seem less forced. Even the mechanics could see that their employers could well be the chief beneficiaries.[69] As one writer put it 'it was only fair to presume, that the better both parties were instructed, the better would the market contract be performed'.[70] Naturally the employers did not stress this point in public, but, even if they were primarily interested in social control, gains of other kinds might make them favourably disposed towards education. It seems safe to conclude, then, that much of the rhetoric of utility was sincerely meant and that it was a genuine expression of strong motivation in the formation of the Mechanics' Institutes. This was shared by employers and employees, at least in the early days, and it is entirely consistent with an ideology that owes much to religious Dissent, political radicalism and the Scottish educational system. But this was not the only motivation. The promoters of science education may have had another rationale.

9.4 THE MYTHOLOGY OF SOCIAL CONTROL

The thought that the employers who promoted science education might have devious intentions that they were reluctant to declare occurred to several of their contemporaries. On the one hand, the fears of 'a country gentleman' that workers 'puffed up with pride of science' would lead to an English re-run of the French Revolution[71] were accompanied by nasty hints that 'Brougham wishes, perhaps, for merely political purposes, to count a noisy mob on his side'.[72] On the other hand, the suspicion that the masters were up to something reinforced the demand for worker control of the Mechanics' Institutes and led to such breakaway movements as the Manchester New Mechanics' Institution, Glasgow Mechanics' Institution, The Halls of Science and so on. In a curious way both right and left seemed to fear that the Mechanics' Institutes would be used to effect social changes that did not depend directly upon the application of useful science. In recent times the theme has attracted considerable attention,[73] but 'social control' has tended to become so dominant in some writings that other, equally relevant, considerations are excluded. Perhaps because of its politically evocative overtones it has in fact developed a mythology of its own. So it may be helpful to identify the elements in the proposition that science was intentionally used in the Mechanics' Institutes as a form of social control. At the very least it must involve these beliefs: (a) that education can be a means of social control; (b) that science itself was particularly appropriate for this purpose; (c) that science would need to be conceived and taught in a certain kind of way so as to achieve this objective.

It has been widely agreed by historians of education that, in this period, education was quite generally seen in this light.[74] The fact that philanthropic, altruistic motives can also be recognised does not contradict the thesis that education could be, and was, used to effect social control in a particularly turbulent period of English history. In the case of the Mechanics' Institutes, one needs to see first how such control could be envisaged in a quite general way, i.e. independent of subject matter taught.

One obvious method was by combating sensuality. It was often assumed that workers were specially prone to 'sensual' as opposed to 'rational' diversions, so a Mechanics' Institute provided them with 'pursuits above the grossness of sensuality' and 'reclaimed many from the habits of vice. It provided them with safe and rational recreation, which might otherwise be sought in scenes of low debauchery, and it had the effect of promoting the strength and prosperity of the country in general.'[75] Similar sentiments were expressed by Birkbeck and many others.[76]

If sensuality were replaced by sobriety even greater social harmony was likely to prevail. Learning was frequently urged as a substitute for liquor. The Rotherham Mechanics' Institute originated in the local Temperance Society (1838),[77] and one of the founders of the Preston Institution for the Diffusion of Knowledge (1828) was Joseph Livesey, three years later to become founder of British Teetotalism.[78] But often the pursuit of knowledge was explicitly urged as making a man 'superior to habits of drinking'.[79]

Crime was also thought to be displaceable by learning; indeed Professor James Pillans thought 'a well-digested system of national education skilfully carried into execution, would in the course of a generation or two almost extirpate crime',[80] and that ultimate of evils, social disintegration, would be kept at bay by an education which could provide answers to the questioning militancy of an insurgent working class. So thought another Baptist Minister, the Rev. John Foster,[81] and many others like him. With hindsight we may smile at the quaint naïvety of such hopes. But we cannot deny that in the context of the Mechanics' Institutes they were strongly and seriously entertained.

Although this melioristic view of education was widely held in the 1820s and the 1830s it was by no means universal. On the one hand were the Tories like the anonymous 'country gentleman'. On the other were the radicals like Robertson and Hodgskin, and the artisans themselves, who had no such desire to maintain the status quo. So the best we can say is that some, possibly the majority, of promoters saw their objectives primarily in these terms. In any case, it is clear that the audience was not intended to be wholely working class, still less composed of those with revolutionary aspirations.

Again, even if at this very general level, the 'real' aim was for social control, how could the projectors have been so inept as to think that *libraries* would have the slightest effect on the mass of the working populace? Yet these, more than lectures, were often the chief medium of 'instruction'.[82] What working man, left to himself with an assortment of uplifting books (often cast off by their previous owners) would become sufficiently spellbound as to abandon forever his revolutionary conspiracies, his Saturday night carousals, or even his pint?

Some intention of using the Mechanics' Institutes as a social instrument is unquestionable. But in case after case for which data are available there is no evidence whatever that social control was chiefly in the minds of the projectors. To assume otherwise is to create a new myth, which like all myths, has a kernel of truth but a lot of imagination.

The second part of the thesis suggests that science had a very specific function. Clearly there must be some reason for the initial concentration on science, emphasised by the widespread exclusion of religion and politics, and suggesting that there must have been something very special about the role that science was expected to play. If the rhetoric of utilitarianism is rejected, the obvious alternative is that science might have a distinctive contribution to make to achieving a measure of social control. In other words, science could function as a powerful ideological tool. There are at least two ways in which this could happen. The first of these is the very traditional one, by natural theology. By detailed study of the physical universe one could rise 'through nature to nature's God'. For a mechanic who, as we have seen, was commonly supposed to respond more to sensual than to rational stimuli nature could be a more eloquent preacher than those worthies who filled the pulpits of the day. It was not so much the familiar argument from design that was expected to appeal to the artisan. Rather it was thought that the very concreteness of natural phenomena might replace the theological abstractions that were supposed to be beyond his comprehension. Unacceptable social behaviour could then be seen as 'unnatural' instead of just 'immoral'.

Thus, although works of polemic theology were generally excluded from their libraries, mechanics could expect to find plenty of works of natural theology on their shelves, often including a new issue of Paley's classic *Natural Theology*, edited and introduced by Brougham himself. And in those lectures and addresses which have survived, references to the beneficence of the Creator and to the orderliness of creation are legion. This emphasis on natural theology may, however, have been a double-edged weapon, not just because of the difficulties it presented to the artisan (of which we know very little),[83] but for a reason perceived by Lord Brougham himself. He recognised that some were 'alarmed lest the progress of natural religion should prove dangerous to the acceptance of revealed'.[84] If it is true that the working classes were responding strongly to the forces of the evangelical revival,[85] an eviscerated natural religion could strongly diminish the appeal of doctrinal Christianity with its powerful social conscience. This may be the reason why some observers criticised the Mechanics' Institutes for not drawing appropriate moral and religious lessons from natural knowledge.[86] These facts make it difficult to believe that science was favoured because of its possible teleological uses.[87]

Science, however, could display other characteristics than 'the evidence for design'. It has often been thought to show true objectivity, freedom from value-laden concepts, and a general air of factual solidarity, which impart to it a number of desirable social functions. For instance, it could have, for these reasons, a *stabilising* role. Shapin and Barnes stress the dependence of its credibility upon 'studied disinterest and apparent objectivity', and suggest that 'value-free' science, once accepted, could minimise the danger of moral coercion by many kinds of political appeal.[88] Or science could have an *authoritarian* role, compelling obedience by its unchallengeable rectitude, making the recipients at least submissive to their instructors in their science classes, and that was always a useful start. Or again it could have a *mediating* role, binding together people of different religious and political beliefs,[89] and in conformity with the banning of religious and political discussion, provide a common platform of unity. If it were thought that this might paper over the cracks between the different political parties, it was not very successful, since apparently only two Tories supported the project in Manchester (Williams and Birley), only one in Liverpool (the ill-starred Huskisson), and none at all in London. But failure in the event does not necessarily deny the intentions.

Value-transcendent objective science could have been espoused for these reasons; but was this actually the case? While it is easy to dogmatise, the nature of the historical data suggests caution. In a reaction against the totalitarian claims of a positivistic scientism it has become fashionable in our day to deny, even in principle, the possibility of value-free knowledge. A determination to apply this dogma to science on every possible occasion has, in some quarters, become so obsessive that a balanced historical view is almost impossible. If, in fact, projectors of the Mechanics' Institutes thought in these terms about science then unequivocal evidence should be produced for their intentions. This does not yet appear to have been done, which is not surprising since the sophisticated view of science proposed is essentially a modern one.[90]

If this view of science had been held by the promoters, it should be reflected in their curricula and the way in which science was taught. Shapin and Barnes have gone as far as to assert that the kind of scientific knowledge conveyed was 'designed precisely to constrain its recipients and stultify their imagination'.[91] But no statements they cite, indisputably justify this stinging accusation. A suggested preponderance of physics and mathematics (supposed to be the least value-laden subjects) is not borne out either by inspection of curricula or examination of library lists,[92] where, in almost every case, chemistry exceeds physics, on average by a factor of over two to one. Chemistry in its confused state during the 1820s and 1830s was one of the worst examples of a science in which a theory was unanimously held as 'fact'. Exceeding chemistry in importance was engineering which, with its 'margins of error' and (in places) absence of adequate theory, was hardly a paradigm of value-free science. Even where physics and mathematics do play a prominent place in the curriculum, they are the most obvious prerequisites to a study of the principles of engineering, and could be justified in those terms alone.

It is of course perfectly true that, in the presentation of science, simplifying steps were taken by Brougham and others to minimise the theoretical content. Their science was 'hard, factual, solid and enduring; in no way tentative or revisable'.[93] It is true that elaborate theories may be eliminated from a teaching programme because they are embarrassing and revealing; but it may also be because *they are irrelevant to the immediate programme, or simply difficult*. In the search for subtle ideological explanations it is possible to overlook altogether this most obvious and elementary aspect of pedagogic practice.[94]

There are many other explanations possible for a diminution in theory. At Sheffield, for example, a lack of theory in chemical lectures reflected the current state of chemistry and variety of experience of the lecturers.[95]

A third characteristic of science that could contribute to its effectiveness in social manipulation is its co-operative nature:

> Meeting, as both classes do, on the fair field of science, where all are as brothers, and pursuing, it may be, the same glorious objects, the wall of separation is removed for ever, and the best possible guarantee given for the inviolable maintenance of the rights of property on the one hand, and the peace and security of society on the other.

This splendid citation, for which we are indebted to Shapin and Barnes, is not however from one of the projectors. Not surprisingly, this egalitarian view of science was that of a humble member of the Glasgow Mechanics' Institute, from a prize-winning essay.[96] There does not seem to be much evidence that the projectors saw – or wanted to see – such a levelling effect of science. They certainly wanted social harmony, but to achieve it they tended to rely on social events rather than science.[97]

At a different level this sharing of a scientific enterprise, rather than any philosophical views of 'objective science', was the point being made by a Whig director of the Edinburgh School of Arts when he suggested that men of all parties 'could unite cordially in prosecuting a common object', because the common object was, for other reasons, well esteemed by all parties.[98] It is suggested that the esteem of science by artisan and projector stemmed primarily from a sense of its utility (real or imagined) and also from local considerations that we shall now briefly examine.[99]

9.5 THE SYMBOLISM OF PROGRESS

The massive employment of science in both the propaganda and the practice of early Mechanics' Institutes is not to be simply explained in terms of technical utility (as some of the promoters claimed) or of the power to control the working classes (as some historians suggest). One other characteristic of science that made it eminently appropriate for individuals and communities to adopt it as their emblem was its progressive character. Far from having a static view of science, many of the movement's leaders were vociferous over the extent to which it had changed, even if its progress was by finely-graded steps rather than sudden revolutions. For rather obvious reasons the latter were fairly unpopular in the 1820s and 1830s, and a belief set in of steady, if not inexorable, progress. It will be recalled that most, though not all, of the leading lights of the movement were Whigs, and their view of science was an exceedingly Whiggish one. Here indeed there is a strong and close association between science and society but not, it would seem, one that was in any way coercive.

Perhaps the finest statement of this attitude was by Brougham himself when, late in life, he unveiled the new statue of Sir Isaac Newton at Grantham (1859). Observing that the discoveries of Newton were subject to the law of continuity, of gradual progress, 'which governs all human approaches to perfection', he went on to claim 'nor is the great law of Gradual Progress confined to the physical sciences; in the moral it equally governs'. As an illustration he suggested that this law 'allows "Mixed Monarchy" to be established, combining freedom with order – a plan pronounced by Statesmen and writers of antiquity to be of hardly possible formation, and wholly impossible continuance'.[100]

Indeed, on occasions when social progress appeared to be slow or non-existent Brougham would actually take heart from the progress of science. He tells us how, following the gloom which descended upon the Whigs after the General Election of 1837, he and a companion regarded the current political scene as 'a subject less suited to engage our conversation; and we naturally dwelt little upon passing and unpleasing topics'. And so they turned to 'those matters of permanent interest and universal importance, and which the follies or faults of men could not despoil of their dignity or deprive of their relish' – that is to the study of nature.[101]

Science was thus a potent, discernible symbol of change in general, including the hoped-for changes in society. And perhaps because it still had a certain novelty, as well as the reputation for having brought about practical good, it was desirable that all forward-looking progressive people should espouse its cause, and, given the existing movements for popular education, it should be taught to all and sundry. In support of this view may be cited the otherwise surprising emphasis to be found in many science books at that time on the history of science,[102] and frequent triumphalist references to its progressive character.

The role of science in English provincial culture is a complex one and needs further examination. However recent studies on Manchester,[103] Sheffield,[104] Derby,[105] and the Potteries[106] have firmly located the development of the Mechanics' Institutions within the local culture, and as an expression of social aspirations in provincial centres. Time after time many of the leading projectors were local politicians who sought to maintain the status of their town in the surrounding region. Such was the case with Christopher Johnson of Lancaster who, while Mayor, fought strongly against the removal of the Assizes from the County Town.[107] He it was who encouraged the young Edward Frankland and his friends in their early encounters with science. A recent study of the movement for adult education in Warrington concludes that 'the part played by local pride . . . is a factor that cannot be over-looked'.[108] This spirit of local competitiveness ran high; in 1845 newsrooms were hastily provided when it was discovered that they were common elsewhere; in 1857 a humiliating recognition that 'our Institute will not bear comparison with many others' was an immediate catalyst for further reform. In general it was concluded that 'a desire to promote a community spirit was connected as much with a growing civic pride as with any policy of social control'[109] and, as Inkster has reminded us,[110] local factors extend considerably beyond hegemony within a region. Local political alliances, institutional networks, industrial practice and general economic prosperity all played a part in fashioning the use of science in Mechanics' Institutes. Finally, it must not be forgotten that here was an opportunity for individuals to exercise creative leadership. Whether or not these were the 'marginal men' supposed to be finding release from their marginality in the Lit. and Phil. Movement (and in some cases they possibly were), they were often people whose non-conformity debarred them from cultural leadership in church and university, and who genuinely wanted to use their energies in a manner both progressive and philanthropic. As Joseph Livesey of Preston admitted 'I had a restless spirit, and was always projecting something new. After seeing an institution fairly and successfully started I began to feel indifferent'.[111] Further south another zealot for popular sicence education demanded 'may we not indulge the hope that Manchester will one day rank as high in the literary and scientific world as it now does in a commercial point of view?'[112]

Without any doubt the early Mechanics' Institute movement is an excellent illustration of 'the band-wagon effect'. Once the diffusion of science among working people had been taken up, for good reasons or bad, others simply had to follow suit.[113] Its decline as social conditions changed became a national, rather than a local, phenomenon. Until that time the available evidence suggests the role of science in the Mechanics' Institutes was chiefly to minister to local patriotism, frequently to provide useful practical knowledge, but only occasionally to serve as an instrument for controlling the turbulent society of England in the early nineteenth century.

10. Strongholds of Amateur Science in England

10.1 'NOT THE LAND OF SCIENCE'?

IN 1830 a remarkable work issued from the pen of Charles Babbage Esq., FRS, Lucasian Professor of Mathematics at the University of Cambridge (and thus distant successor to Sir Isaac Newton), computing pioneer, and seeker after an infallible method for predicting the winners of horse races. The book in question concerns neither the refinements of his latest 'calculating engine' nor the intricacies of punting. It was an expression of acute despair entitled *Reflections on the Decline of Science in England.* He concluded 'the pursuit of science does not, in England, constitute a distinct profession, as it does in many other countries', and so, 'when a situation, requiring for the proper fulfilment of its duties considerable scientific attainments, is vacant, it becomes necessary to select from among amateurs'.[1] How far Babbage was justified in his overall analysis of a 'neglected and declining' science in England, and in his particular diatribes against the Royal Society, has been much debated of late and is perhaps no longer very important, since his evidence was selective and ignored almost entirely provincial science. Probably the book is as much an expression of personal frustration as anything else. But in his castigation of much of English science as 'amateur' he was undoubtedly correct. Liebig was even less flattering in 1837: 'England is not the land of science; there is only a widely dispersed amateurishness'.[2]

Even as Liebig was writing, however, events were beginning to prove him wrong. Invisible perhaps to the jaundiced eye of the Cambridge don, and beneath the notice of the great representative of German science, they were the first steps towards the professionalisation of science in England. It was to be a long haul, and indeed it is not complete even yet. During the early stages the emerging professional and the established amateur could happily coexist side by side. But the writing was on the wall, and the long English tradition of science as an almost entirely amateur pursuit was coming to an end. This decline of the amateur, just as the so-called 'decline of science', meant, however, not annihilation but metamorphosis. It meant not a decline in overall numbers, but a decrease in the area of territory over which the amateurs held undisputed sway. Nothing can illustrate the change more clearly than the histories of those institutions which were the haunt of the amateur men (and women) of science. Such people, because they had the time and money to indulge their scientific inclinations, were almost always from the middle or upper classes at first. But here as elsewhere changing social patterns were reinforced by the changing nature of the scientific enterprise. As is so often the case the changing character of institutionalised science both reflected and induced further changes in the wider social context.

10.2 THE 'LIT. AND PHIL.' MOVEMENT

The Lit. and Phil. movement, as has been seen, originated at the end of the eighteenth century. Nearly always a middle-class enterprise, it enshrined certain cultural attitudes towards science, and these it attempted to preserve and propagate within the culture of the locality. To belong to a Literary and Philosophical Society was a status symbol in itself, partly because of the high esteem in which science was held by the community. That, at least, was the theory many of the promoters seemed to hold. Judging by lecture courses, library lists and announcements by the founders themselves, science was a major raison d'être for the Societies' existence, though we may be permitted to suppose that this was true for a variety of reasons, not all of which were held with the same force. It is all too easy to generalise.

In the course of time each Society tended to develop in its own characteristic way. By the end of the Napoleonic wars a second or even third generation was populating the original Societies, and older values did not necessarily pass unquestioned. Crucial to the development of these Societies was their developing conviction of the role which science ought to play in their affairs.

In towns which had no Lit. and Phil. earnest citizens sought to rectify the deficiency with all speed. Institutions of this kind sprang up at Liverpool (1812), Portsmouth (1818), Leeds and Hanley (1819), York, Hull and Sheffield (1822) and in many other centres. As with the Mechanics' Institutes one compelling reason was the need to minister to civic pride. As local conditions varied, so did the composition and pattern of the new Societies. At Liverpool, the sixteen founders were by no means all 'marginal men', but included merchants and gentry and an Anglican clergyman. However over half were nonconformists, and five had held politically radical views. There were three physicians or surgeons.[3]

On the other hand the Leeds Philosophical and Literary Society numbered seven medical men amongst its twenty-one first members in 1818, and had a roughly equal balance of liberals and conservatives, politically and ecclesiastically, no less than six former mayors were included; but only three manufacturers, suggesting a greater awareness of civic dignity than the usefulness of science.[4] Likewise at Swansea (1835) it was suggested that, since comparable societies existed at Bristol and Neath, Swansea ought to follow their lead.[5]

Although differences of nomenclature are notoriously misleading, it is significant that the period following the French revolutionary wars saw the emergence of several 'Literary and Scientific Institutes'. Three of these appeared in London in the 1820s, intended 'for persons engaged in commercial and professional pursuits'. The middle-class equivalent of the Mechanics' Institutes, they tended in due course to be indistinguishable from the latter, and were mainly to be found in the South of England.[6] The novelty of their title stresses the importance attached to science by their founders, and it was this growing recognition of the social relevance of scientific knowledge that constituted one half of the challenge faced by all such institutions, old and new. The other half was an insistent desire for 'clubbability', a recognition of social needs transcending the attractions of natural knowledge, and, as the latter became increasingly specialised and 'difficult', it became correspondingly harder to reconcile the two needs. How this tension was resolved depended entirely upon local circumstances.

At one extreme it was possible to play science down to such an extent that it scarcely featured in the programme or title of the institution. This was the case in a few small towns in the rural heartland of England. Bedford, for example, gave birth in 1826 to a 'Reading Room Society', which commenced its existence by ordering the *Quarterly* and *Edinburgh Reviews*. The first books recorded in this library were a copy of Linnaeus, one of Gilibert's *Fundamentorum Botanicorum pars Prima* and an edition of *Don Quixote*.[7] The town did not have a Literary and Scientific Institute until November 1846, and although that did run some classes in chemistry it appeared more concerned to provide musical entertainment. In the face of competition from a Working Men's Institute (founded 1855), it revived its flagging fortunes by amalgamation with the Reading Room Society, this maintaining a strong conservative and anti-scientific tradition.[8]

Many post-war Lit. and Phil. Societies started with laudable intentions of disseminating science. Two very different institutions, both founded in 1835, were united in this respect. At Swansea, science lectures predominated at the Phil. and Lit. (as it was curiously called at first), whose founders included three Fellows of the Royal Society, the President of the Geological Society (de la Beche) and numerous manufacturers.[9] At Sunderland, on the other hand, the founders were lesser known local men, the patron being the Earl of Durham. Its first three years' programme included lectures in chemistry, electrical science and (perhaps inevitably) phrenology and natural theology.[10] Though the Swansea Society managed to survive – it even became the Royal Institution of South Wales – its less well-endowed counterpart in Sunderland continued only by merging with the Mechanics' Institute. Just as was the case with many literary and scientific institutes the ideology of improvement for the working class proved more powerful than self-advancement for the middle class, in a science that was increasingly difficult and for which the cultural value was manifestly diminishing. If, as many have argued, there was an element of social self-justification through science for the magnates of the early Industrial Revolution, this need was less obvious in a society where their children and grandchildren felt much more secure. By 1835 other societies were disappearing from view. The Potteries Philosophical Society expired that year, unable to cope with competition from the local Mechanics' Institute, which lured away its lecturers and even its Patron. Again the same forces were at work, though, as Shapin observes, a certain scale of population, with appropriate social diversification, was necessary for success. In places like Hanley there were simply not enough people with the necessary ideological commitment to sustain an exclusive and relatively élitist institution.[11]

Merger with, or replacement by, Mechanics' Institutes became a dangerously possible fate for a Literary and Philosophical Society. But other possibilities existed. Another example, diametrically opposed to that of Bedford, was the extreme case of Manchester. The Manchester Literary and Philosophical Society[12] was unusual in that the leadership had changed very slowly over its past sixty years. Thus John Dalton, founder of the chemical Atomic Theory, was President for twenty seven years until his death in 1844, when he was succeeded by Edward Holmes, whose Vice-Presidency stretched back to 1798! And it was certainly unique in having such a world famous man of science as its leading light for so long. When Dalton died he was given a civic funeral, the procession stretched for a mile and all the shops were closed. That was the measure of Manchester's pride in her most famous citizen; it does not imply that most people believed, or even understood, the theory he had propounded.

It was perhaps Dalton's influence and importance that prevented stagnation, for the amateur membership and cultural comprehension of science changed little until the 1820s. Then changes began to be felt. New men of business wanted to diffuse useful knowledge among the masses, and the Mechanics' Institute was born. But that did not threaten the Lit. and Phil. for a new breed of scientifically aware gentlemen and manufacturers was determined to make it into a clearing house for technical information. There was little of the polite interest in science as a genteel activity about the engineers Fairbairn, Hodgkinson, Osborne Reynolds and others. When, in the late 1840s, they assumed leadership it was, as Kargon has aptly called it, 'the changing of the guard'. Over the issue of whether it should include in the *Memoirs*, as a tribute to its late President, his posthumous paper on 'The History of Sculpture', the Society was divided. In the end the essay was printed, but from then on ninety-five per cent of the papers were on scientific or technical matters, reflecting a new resolve to turn the Society into a major instrument for scientific advancement and research. The *Memoirs* in fact became an internationally famous journal.

This option of concentrating almost exclusively upon science was open to few other amateur Lit. and Phil. Societies. Nowhere else in England outside London was there such a thriving scientific community as in Manchester, nowhere such a sense of civic pride in science, and nowhere such continuity of tradition combined with a juxtaposition of innovatory scientific and technological interests. Yet less fortunate societies could still survive by a combination of judiciously blended objectives and a series of happy accidents. Such a case was that of Newcastle. Its engineering and chemical concerns developed rather later than in Manchester, though the Literary and Philosophical Society (founded 1793) played a significant role in the early lives of George and Robert Stephenson. As at Manchester, by the 1830s the old style enthusiasm for science was beginning to falter, but at Newcastle the responses were quite different. As far back as 1814 concern had been expressed by the Managers about the shortage of papers for monthly meetings. When a Committee was appointed to consider publishing those that had been presented, its excuse for doing nothing (1831) was distraction by the current political crisis.[13] Over the first thirty years of the century some extremely important discoveries had been conveyed to the Society, including George Stephenson's safety lamp and the world-famous process devised by H. L. Pattinson for separating lead and silver. But the 1840s was a time of doldrums and from 1856 there is no further reference to monthly meetings. Meanwhile the other main academic activity, the Lecture Programme, shows similar signs of change with, in 1830, the first Lecture Course on non-scientific matters: six lectures on 'Origins and Progress of Civil Society' by the Rev. William Turner Jnr., son of the founder and significantly representative of the new generation. Thereafter the Lecture Programme continued with an uneasy and irregular mix of science and non-science which appears to indicate no consistent policy and was perhaps dictated (especially in the lean years) by whoever happened to be available (see figure 5). Ideological ambivalence was reflected

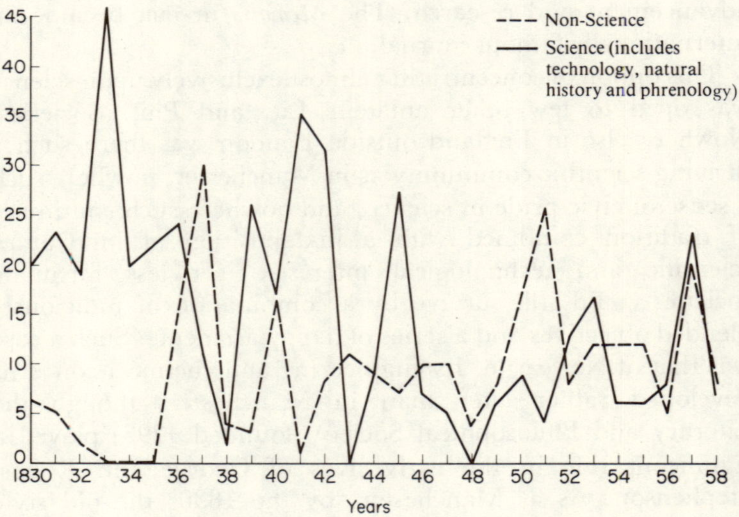

FIGURE 5 Lectures to Newcastle Lit. and Phil., 1830–1858.

in financial instability and by 1849 the Society was seriously in debt.[14] A bazaar and an appeal were both singularly unsuccessful, but after much squabbling the situation was saved at the eleventh hour by a gift, and shortly afterwards a legacy, from Robert Stephenson. The annual subscription was reduced and by the next Annual Meeting (1857), membership had more than doubled from 506 to 1016. The Society had been saved, and it continues to this day, but its role as a major exponent of provincial science was over.

In 1851 J. W. Hudson exclaimed 'the Provincial Philosophical Societies of England have completed their career, they are the debris of an age passed away'.[15] And so they were, at least in their original conception (Manchester uniquely excepted). The apologist of the Newcastle Lit. and Phil., R. S. Watson, explained the demise of the monthly scientific meetings 'partly from the constant increase in the number of local societies each with its own specific object of research, and partly from the rapid extension of the railway system'.[16] For a city so remote from London his second reason may seem credible. However rail connection between London and Tyneside was not completed until 1844 (not until 1850 for Newcastle itself), and the swing from science had begun many years before. More relevant is his first reason: the formation of new specialist societies. Amongst these were the Natural History Society (1829), the Tyneside Naturalists' Field Club (1846), The North of England Institute of Mining Engineers (1852) and the Newcastle Chemical Society (1868). Although their existence eventually spelt death to any hopes of the parent body becoming an active research institution, by syphoning off the specialist talent that had been accumulating, they eliminated a potent cause for internal strife between old and new style members, and probably meant that the Newcastle Lit. and Phil., unlike so many contemporaries, could continue and even flourish in a rather different role.[17]

10.3 THE NATURAL HISTORY MOVEMENT

From the decline in science experienced by the Literary and Philosophical Societies one is easily tempted to infer, in the manner of a Babbage, a decline in provincial amateur science generally. In so far as this applied to chemistry, physics etc., it is not far from the mark. But modern science has inherited the traditions of natural history as well as of natural philosophy, and it was in the region of natural history that amateur science survived, and in fact flourished mightily in many provincial centres of Great Britain during the nineteenth century. That this often happened outside the ambit of the Lit. and Phil. Societies reflects their own inability to support the specialist research being demanded.

From the beginning of the nineteenth century the cult of the naturalist was a growing one in Britain.[18] At a deep ideological level it was a response to the Romantic appreciation of nature that was otherwise and variously expressed in the chemistry of Humphry Davy, the continual appreciation of mountain scenery and the poems of William Wordsworth. Not surprisingly it was strongly felt in the most industrialised urban communities. Lynn Barber has argued that the real reason for the rise of natural history in the early nineteenth century was the need for an antidote to boredom, and that its unique status derived from the dominance of natural theology.[19] While these explanations may be partly true for amateur science in general, they are quite inadequate to account for the specific shift of emphasis away from chemistry and physics towards the field sciences. So we must look elsewhere.

Followers of natural history tend to report their findings at great length and with prolific illustration, and so they received specific encouragement by the new availability of cheaper books and journals. These were made possible by the advent of steam driven printing presses, and the reduction (1836) and later repeal (1861) of taxes on paper. Another fiscal measure favouring natural history was the repeal of the glass tax in 1845; this not only made possible such spectacular structures as the Crystal Palace (1851) but also the multitudes of glass cases and aquaria which swept into fashion at exactly the same time.[20] The great improvement in travel, especially by coach, gave a wider dimension to natural history than might be suggested by the solemn publication of lists of all insects netted during single coach journeys.[21] The beginnings of mass migration to the seaside later added to the possibilities of collecting marine objects from the seashore. And the development of field equipment for medical men, who were compelled under the Apothecaries' Act of 1815 to study botany for purposes of herbalising, led to its adoption by a wider, amateur audience.

The various crazes for entomology in the 1820s, for fern collection in the 1850s, and so on, did not necessarily have anything to do with scientific enquiry. To fill one's house with 'nature' might betoken a sentimental reaction against urbanisation, the mindless following of fashion, or mere acquisitiveness made easier by the general rise in prosperity in the 1850s. Some, indeed, of these attitudes were anti-scientific, as when the despoliation of the countryside endangered or eliminated rare species of flora and fauna.

But the very abundance of 'objects of nature' in Victorian parlours meant at least an awareness of their rich variety, and may well have encouraged a more objective, detached approach to them. Without doubt a genuine spirit of scientific enquiry moved in many naturalists whose collecting instincts had been brought properly under control. These men were nearly all amateurs, and such was the state of the art that an amateur could very well pursue it. Mathematical expertise was unnecessary; experimental technique, though important, could be learned on one's own; no lengthy laboratory courses were necessary; and the intellectual demands, though considerable, were not impossible to meet. The feverish pursuit of natural history occupied many a lone amateur as he burned the midnight oil preparing, classifying, mounting his specimens to his own satisfaction.

There was also much corporate activity, and its locus was the bevy of natural history societies that sprang into existence in the early nineteenth century, and the field clubs with which they were closely related. Thus the Newcastle Natural History Society (properly called The Natural History Society of Northumberland and Durham) was founded in 1829 and took over the collection of natural objects owned by the Lit. and Phil., also using that Society's lecture room for a time, thereby providing a pleasing demonstration of the symbiotic relationship between the new society and its parent. Like many another provincial counterpart it eventually acquired its own premises containing museum, library and lecture room, much in the manner of the Lit. and Phils. Such societies tended to be fairly exclusive, with high subscriptions as a price for the prestigious accommodation and often the learned journal as well. Enormous amounts of information about local plants and animals were published by the larger societies, much of which was of permanent scientific value. The *Transactions* of the Newcastle Society and its related Field Club carried, in the century 1829–1929, reports of one order, one family, 49 genera and 218 species, all claimed to be new to science.[22]

One of the first activities of the Natural History Society was the mapping of the North Eastern Coalfield,[23] a task the Lit. and Phil. had lamentably failed to undertake. Later, in conjunction with the Field Club and the British Association, from 1861 to 1863 it carried out a series of dredging operations in the North Sea, adding considerably to the knowledge of oceanography in the neighbourhood of the Dogger Bank.[24]

The élitism of most Natural History Societies was their undoing. Without a large membership both ideas and money were in short supply, the latter being particularly serious if expensive property had to be paid for and maintained. Thus we find the *Transactions* of the Newcastle Society discontinued after 1838. That was the very year in which the Manchester Society for the Promotion of Natural History (founded 1821) opened its museum to the public. But there was another side to its difficulties. Once the Manchester Society became less élite the upper middle class ceased to support it;[25] eventually its collection was taken over by Owens' College.

If the Natural History Societies were victims of rigid class attitudes (as was often the case in larger industrialised communities) an alternative approach was being worked out in a very different social context, the border country between England and Scotland. It was here, in 1831, that the Berwickshire Naturalists' Club was formed, an institution which owned no property, which opened membership widely to members of both sexes, and whose activities were far removed from the evening sessions of the so-called 'closet' naturalists. After suitable refreshment at a local hostelry early in the morning, members would spend whole days on the bracing uplands of the Cheviot or surrounding countryside, combining the sheer pleasure of open air exercise in congenial company with a serious attempt to study the natural history of the area and record it in a systematic manner.

Admittedly the social changes of the Industrial Revolution had largely passed Berwickshire by, but its Club became the model for hosts of Field Clubs which sprang up all over Britain. In Newcastle the Tyneside Naturalists' Field Club was founded in 1846 under the direct stimulus of the Berwickshire experience.[26] It was largely a fission product of the Natural History Society of the town. With fewer expenses, the Field Clubs gained greatly in popularity and though larger than the Natural History Societies that gave them birth they often retained friendly links and dual membership. The Tyneside Club, now four times larger that the parent Society, enabled publications of *Transactions* to be resumed at once, although now on a joint basis. In other cases the union of younger and older societies gave yet newer institutions of great vigour.

One consequence of these changes, though hardly one that would have been commonly intended, was a measure of social harmony which could not have been achieved in the old exclusive, property-owning societies. At Warrington, for instance, it was noted that the Field Club (founded rather late, in the 1880s) seemed to have a larger upper working-class membership than other societies.[27] The Field Club actually achieved what the Mechanics' Institutes have sometimes been supposed to have sought: a measure of social harmony and integration. Even then, however, David Allen has observed that, when the Field Clubs had an antiquarian section, that was usually populated by Tories, while the naturalists tended to be more Liberal in politics.[28]

As the members of the new Bedfordshire Natural History Society were told by their Honorary Secretary in 1875, 'it is one of the chief advantages of natural history societies that they bring people of congenial tastes into personal relations'.[29] The socially unifying role of natural history was remarkably exemplified by Charles Kingsley. When he became Canon of Chester in 1870 he was anxious to bring together Cathedral and town. An effective way of doing this was a small botany class which he started, and which within a year had become the nucleus of the Chester Natural Science Society. This is how he recalled the Field Trips, with anything up to a hundred participants:

> Those were bright afternoons, all classes mingling together; people who had lived next door to each other in Chester for years, perhaps without exchanging a word, now met on equal and friendly terms, in pursuit of one ennobling object, and found themselves all travelling in second-class carriages together without distinction of rank or position.[30]

It is of course easy to exaggerate the social effects of these pursuits after scientific knowledge. But in view of the very large number of Natural History Societies it is not unlikely that they made a definite, if modest, contribution to social harmony in Victorian England. In many cases, as at Chester, they certainly succeeded in involving both sexes in a manner that was, for its time, truly remarkable.

Of the strength of the Natural History Movement in the nineteenth century there can be no doubt. In 1873 there were no less than 104 Field Clubs reported out of a total of 169 local scientific societies altogether in Great Britain and Ireland.[31] In an age where the 'hard' sciences were becoming more and more specialised here still was a place for the enthusiastic amateur. The other sciences were fast changing, but not, seemingly, natural history, and, as Barber suggests,[32] it is always easier for lay people to follow subjects that are not currently undergoing revolutions. Even the coming of Darwinism did not seriously undermine their confidence, for had not Darwin himself come to his conclusions on the basis of countless careful observations made in the outdoor world? It is significant that one of the few places where institutional natural history seems to have been under some difficulty was Manchester. And this also was one of the very few where the Lit. and Phil. had come to terms with specialism and embraced it. The social needs of the scientific specialist were still being met by the old established Lit. and Phil.

10.4 THE BRITISH ASSOCIATION

Words change their meaning in a curious way. In the 1820s and 1830s 'association' had evocative overtones almost entirely absent today. Although sometimes applied to loose federations formed for their own protection by employers, it was chiefly used in connection with the new trades unions, struggling to gain 'freedom of association', and then, following the repeal of the Combination Acts in 1824/5 to implement it to the full. It is not wholly surprising that the proposal for an association of scientists received a cool reception from certain bastions of established thought, ranging from *The Times* ('when shall this monstrous bubble burst?') to the Tractarians who feared atheism, and including Charles Dickens who imagined his 'Mudfog Association for the Advancement of Everything'. This opposition was prompted by a variety of misgivings, but underlying them all was a fear that such an association might somehow get out of hand, that it might threaten established interests, and above all that it might replace the aristocracy of science with a meritocracy in which new professionals and talented amateurs combined in free and open association. It is this duality of membership and function that has offered a key to the early history and later difficulties of the British Association.[33]

The British Association for the Advancement of Science was formed on 27 September 1831, at a meeting of the Yorkshire Philosophical Society at its Museum in York. Its objects were declared to be:[34]

> To give a stronger impulse and a more systematic direction to scientific enquiry; to promote the intercourse of those who cultivate science in different parts of the British Empire with one another, and with foreign philosophers; to obtain a more general attention to the objects of science and removal of any disadvantages of a public kind which impede it progress.

In order to appreciate the subsequent role of the British Association in promoting the amateur tradition in science it is necessary to recall the composite origins of the Institution. First, of course, was the so-called 'Declinist Movement' associated particularly with the name of Charles Babbage whose book *Reflections on the Decline of Science in England* (1830) has often been celebrated as the immediate cause of the events at the Yorkshire Philosophical Society. Certainly it was in a review of this work in *The Quarterly Review* that Brewster launched his famous appeal:[35]

> Can we behold unmoved the science of England, the vital principle of her arts, struggling for existence, the meek and unarmed victim of political strife? An association of our nobility, clergy, gentry, and philosophers, can alone draw the attention of the sovereign and the nation to this blot upon its fame. Our aristocracy will not decline to resume their proud station as the patrons of genius . . . The prelates of our national church will not refuse to promote that knowledge which is the foundation of pure religion, and those noble inquiries which elevate the mind, and prepare it for its immortal destination.

As we have seen the chief concern of Babbage had been the low state of the Royal Society. He was not alone in lamenting a general 'decline of science' at the time. He cited Herschel's lament that, compared with continental countries, 'we are fast dropping behind', and indicated that no less a person than Sir Humphry Davy had commenced a work with the same title as his own, cut short by his untimely death.[36] Nor were matters improved by the Royal Society's decision to elect as President not Herschel the distinguished astronomer but a non-scientific member of the Royal family: the Duke of Sussex, sixth son of George III.

Both personally and institutionally then, the 'declinists' had a case, but to assume that the origins of the British Association lay chiefly in their complaints is no longer tenable. Other factors now seem to have been at least as important.[37]

The superior condition of German science, regretted by Herschel, had recently received further publicity. In response to a call from the Professor of Natural History at Jena, Lorenz Oken (1779–1851), 'a great yearly meeting of the cultivators of natural science and medicine from all parts of the German fatherland' was inaugurated at Leipzig in 1822: the Deutscher Naturforscher Versammlung. From inauspicious beginnings it met in different centres for the next few years until, by 1828, its meeting at Berlin was crowned with royal favour, an elaborate social programme and well attended scientific meetings of all kinds. Such an institution was particularly appropriate to Germany which was, after all, not one country but an assemblage of states sharing a common language and cultural heritage, but lacking political coherence and any great metropolitan centre. An annual itinerant gathering of this nature proved an ideal opportunity for statehood to take second place to science, and for a massive if temporary, extension of the German tradition of wandering scholars. But not all attenders came from the fatherland. Two Britons who savoured the delights of international scientific fraternity were Babbage (Berlin 1828) and the chemist J. F. W. Johnston (Hamburg 1830). Each described his experiences in the *Edinburgh Journal of Science*, and neither spared his readership the painful contrast with the Royal Society, particularly in its failure to unite royal patrons and cultivators of science in a common endeavour.[38] The 'decline of science' school thus received new and powerful support, though when Brewster wrote in 1831 to John Phillips, Secretary of the Yorkshire Philosophical Society, his proposal for a British Association was explicitly linked not to the vices of the British institution but to the virtues of the German one.

Events in Germany were not the only stimulus in forming the British Association, however. The Scottish scientific tradition (page 89) was in part epitomised by Brewster whose key role has been touched on already. David Brewster (1781–1868) was one of Edinburgh's most distinguished men of science. A Fellow of the Royal Societies of Edinburgh and London, with an impressive record of research in optics, he was also known for his writing and editing activities, especially in connection with the *Edinburgh Journal of Science, Edinburgh Philosophical Review* and the *Philosophical Magazine*. Aware of the rich tradition of Edinburgh science which he had inherited, he was acutely aware also of the small extent to which scientific achievement was matched by material reward. With this background and a personal appreciation of 'the decline of science', he was, as A. D. Orange has observed, in the best position to turn 'a phrase into a campaign'.[39] What he needed was a platform, and one was readily to hand in his own *Edinburgh Journal of Science*. For over a year he used this to supplement the arguments originally published in the *Quarterly*

Review. Unfortunately his campaigning coincided precisely with the final stages of the battle for Parliamentary Reform and literary fulminations were manifestly not enough. Perhaps the most abiding proof of his organisational genius was a decision to 'plug in' his movement to another tradition that could infuse it with sufficient energy and direction to bring the project to fruition.

The tradition was that of the Literary and Philosophical Societies of England. For reasons still not clear, Brewster sought to float his ideas at a meeting of the Yorkshire Philosophical Society, whose Secretary was John Phillips.[40] In the event its President, William Vernon Harcourt (1789–1871) became the principal architect of the new British Association, differing from Brewster in his greater advocacy of scientific communication, mobilisation of amateur talent and directed scientific research. The declinist rhetoric was quietly replaced by more positive declarations of the need for science to play a greater part in national life. Interestingly, neither Babbage nor Herschel was present to orchestrate the old laments in York. Harcourt's view of the scientific community was more catholic than Brewster's, and his demands for a unified association where men of all kinds of scientific interest could meet was far removed from an Edinburgh model that was exactly in tune with the ethos of the Lit. and Phil. movement of northern England.

In one other way was this tradition carried on in the British Association – the welcome given to amateur practitioners of science. Cry as Babbage might for the professionalisation of science, the British Association in its early years echoed with the contributions of amateurs and gave them a national scientific audience possible in no other context. Indeed Roy MacLeod has argued that at this time '*amateurism* not professionalism was the keystone of British attitudes towards science' and that the 'professionalisation model' of Pearce Williams and others[41] is tenable only 'if one excludes the history of the Association after the 1850s'.[42] It is possible to go even further and assert that in its earliest days the amateur interest was prominent. One has merely to glance at membership lists, readily understandable in the light of the only criterion for admission: membership of 'Philosophical

Societies in the British Empire'.[43] Eligibility for service on the General Committee was 'by the publication of works or papers',[44] but since that could include the most simple and unsophisticated essay even the leadership of the Association was open to amateurs with the most modest scientific confidence.[45] As for the papers read, that inveterate American critic of amateurism in the British Association, Joseph Henry, complained that they contained 'not a few of very puerile communications'. But, as he said, this was to be expected 'from a meeting of the kind where every person has the liberty of making a communication and where the merest *Sciolist* and the profound *Savant* are on the same apparent level'.[46]

Underlying the stress on amateur science lay the philosophy of Baconianism that permeated so much of the Lit. and Phil. movement, especially in northern England. There was a place, and an honourable place at that, for a vast body of humble fact-collectors whose discoveries could be incorporated into a grand synthesis compiled by those with time and skill to do so: the academics and the professionals. Indeed the latter could actually give direction to the amateur body,[47] thus exactly fulfilling the intentions of Harcourt who, unlike Brewster, had a strong Baconian commitment. Not all sciences were equally suitable for such an approach, and it is highly significant that the one which could gain most from it was amongst the most prominent in the early meetings: geology.[48] But writing of his visit to the British Association at York in 1844, Liebig told Faraday that it 'did not satisfy me in a scientific point of view. It was properly a feast given to the geologists, the other Sciences serving only to decorate the table'. He added that 'without a thorough knowledge of physics and chemistry, even without mineralogy a man can be a great geologist in England'.[49] The tradition lasted out the century, as in the project for collecting photographic evidence of geological formations, reported from 1890 to 1937.[50] As Philip Lowe reports an annual Conference of Delegates of Corresponding Societies, established in 1885, 'was recognition that the vast army of local amateur naturalists still had a significant, though distinct, role to play in the national scientific enterprise. The organisation of the Conference and its preservation of the amateur ethos in science has imparted a distinct imprint to the development of the field sciences in Britain'.[51]

In conclusion one may note that, as a corollary of its commitment to amateur science, the British Association has always had a missionary zeal for disseminating science in unenlightened places (which is why Whewell opposed a meeting at Cambridge!).[52] The effect of the visit was not only to stimulate improvements to museums, botanic gardens, teaching institutions and the like, but also to raise generally the level of scientific awareness in the community, and thus to add to the volume of amateur practitioners of science. When this mission was successful (and it often was) the diverging interests of amateurs and professionals became increasingly apparent. The subsequent tensions within the British Association then became a microcosm of the larger struggles within British science as a whole.

11. The Rise of the Specialist

11.1 LANGUAGE AND INSTITUTIONS

IT is almost certain that Michael Faraday was not afflicted by a lisp. The many glowing reports of his lectures at the Royal Institution would surely have referred to it. Yet he objected to a word 'both to my mouth and ears so awkward that I think I shall never be able to use it'. The excessively sibilant word complained of, with the 'equivalent of three separate sounds of S', was 'physicist'.[1] His correspondent William Whewell, soon to become Master of Trinity College, Cambridge, had a flair for inventing new scientific words, such as anode, cathode, anion, cation, etc. He had recently suggested 'physicist' as the equivalent of the French *physicien*, meaning someone who 'proceeds upon the ideas of force, matter and the properties of matter'.[2] It was now coming into vogue as the expression of their sense of social identity by a body of scientific practitioners who wished to make it clear that their concerns were not to be confused with those of chemistry, astronomy or any of the other recognised scientific disciplines. Instead of the older 'natural philosophy', their preoccupation was to be narrower and more specialised. The chemists, with their ancient tradition of assaying, alchemy and medicine, had acquired their distinctive name even before the publication of Robert Boyle's *Sceptical Chymist* (1662). Now the men of physics were to take a leaf out of their book. Others had done the same, notably geologists, zoologists and botanists.

Another neologism for which Whewell was responsible was 'scientist'.[3] This, together with earlier attempts like Charles Lyell's 'scientifics', expresses a new corporate consciousness amongst those who pursued science, either as a career or as a dominant and serious leisure activity. It expressed their differentiation from those who pursued other culturally prestigious activities, such as music and the arts. And at the very same time, within science itself, there came a fragmentation into the so-called 'subjects' or 'disciplines'. It is in fact quite impossible to argue for the priority of one or other of these processes of differentiation. Much more work on the perception of 'boundary conditions' between individual sciences is necessary before we can be sure that specialism either preceded or followed the emergence of the 'scientist'. At the moment it looks as though both processes occurred together.

The new specialist consciousness was institutionalised in a relatively large number of societies which sprang up in the early years of the nineteenth century. Together with a few older examples, they include the following, all of which were situated in London:

Medical	1773
Linnean	1788
Mineralogical	1799
Chalcographic	1803
Medical and Chirurgical	1805
Geological	1807
Astronomical	1820
Meteorological	1823
Zoological	1826
Geographical	1830
Entomological	1833
Botanical	1836
Microscopical	1839
Pharmaceutical	1841
Chemical	1841

A Physical Society was not formed in London until 1874. Student societies devoted to chemistry and to medicine appeared in London and in Edinburgh, the latter capital also possessing a Royal Physical Society and a Botanical Society. Natural History Societies of various kinds sprang up in many provincial centres, reflecting specialist interest, but still available to a largely amateur clientele (see p. 183).

Several of the metropolitan societies were, as D. E. Allen has said, part of a 'tangle of minor bodies', catering for the under-class of the scientifically inclined who could never hope for election to one of the major bodies. These 'outsider' organisations included the Mineralogical and Botanical Societies. They tended to be radical in outlook and to admit women – which some of their more prestigious counterparts failed to do for many years.[4] The Geological, Linnean and Chemical Societies opened their doors to women in, respectively, 1904, 1919 and 1920.

11.2 GEOLOGY FOR GENTLEMEN?

The Geological Society of London was founded in 1807. As such it was the first learned society for geology alone. It also had the unenviable distinction of being the first to bring down upon its head the wrath of the President of the Royal Society, Sir Joseph Banks. The Royal Society's objections to this upstart were as much against its independence as its specialism, yet were ironically a tribute to its virility. Its founder and first President was G. B. Greenough (1778–1856). According to him the Society began as a gentlemen's dining club, convened to promote the publishing of a mineralogical treatise by L. Bournon, an aristocratic French emigré. That accomplished, most of the fourteen subscribers continued to meet for pleasant company, together with a few others including Humphry Davy. In November the club turned itself into a Geological Society, with the express purpose of facilitating communication between geologists, unifying nomenclature and identifying areas of future research. Such an enterprise would require a network of members over the whole country, and fourteen 'Honorary

Members' were elected from beyond the London area. A Patron was elected, the Hon. Charles Greville, a wealthy FRS, renowned for his collection of minerals.

The manner in which the newly fledged Geological Society reflected changing social awareness about the utility of geology is a matter for debate. The picture painted by Greenough in his private papers is of a gentlemen's dining-club changing quite quickly to a research-orientated, fact-collecting agency, with all the trimmings of a learned society, possessing its own premises (to house collections), publishing its own *Transactions* and exacting high membership fees in order to pay for both. The pressures to bring it under the wing of the Royal Society were stoutly resisted by the young no-compromise President. Faced with a proposal that its government should be in the hands only of members who were also Fellows of the Royal Society, he objected 'to any other aristocracy than one founded upon merit'.[5] But the Geological Society was not fully a meritocracy. The monthly dinners at fifteen shillings a head were unlikely to appeal to the engineers, miners and other 'practical men' invited to join. The glossy *Transactions* were also too expensive for such people; in any case their first decade of issue included only about nine papers out of eighty-eight with any considerable practical interest,[6] and the Society's ranks did not include several practical men of obvious merit, the most outstanding omission being William Smith (1769–1834). As a surveyor in the Cotswolds, and then in Somerset, he travelled widely in the course of his work, which also included superintendence of canal construction. His study of the strata of different districts led to the crucial recognition that each bed of rock could be identified by the fossils it contained. His vast countrywide experience should have made him invaluable to the Geological Society, yet he never joined. It is possible, as his biographer suggested,[7] that Smith felt a proprietorial interest in stratigraphy and that the Society was viewed as an unwarranted interloper. Be that as it may, the omission of 'the Father of English stratigraphy' from the Geological Society casts some suspicion on the utilitarian rhetoric that marked its early days. Some methods could surely have been found to allay his fears and resentment, if indeed they existed.

Roy Porter has gone considerably further and asserted that, with the exception of Smith, none of the leaders of British geology in the first dozen or so years of the nineteenth century had a significant involvement in the practical arts like mining.[8] Apart from the gentlemen of leisure, there were medical men and a number of academics, including Jameson and Playfair (Edinburgh), Daubeny, Kidd and Buckland (Oxford) and Sedgwick (Cambridge). He also listed several other important figures in practical geology who remained outside the Geological Society, including the mineral surveyors Robert Bakewell (a consultant to the landed gentry) and Smith's associate and protagonist John Farey. In these circumstances Porter feels, not that utility was necessarily a 'false banner' for the Geological Society, but rather that it was a concept more comprehensive than mere technical applicability, embracing 'intellectual, religious and moral benefits quite as powerfully as material advantage'.[9] In other words, it had the same broad cultural overtones as in the contemporary Literary and Philosophical Societies, only here it was confined to a narrower and deeper channel, that of one specialised science. This reading is consistent with Greenough's admission of the polite origins of his Society and even with its early utilitarian propaganda. Perhaps Buckland put it as well as anyone in a citation by Portock in his Presidential Address to the Geological Society that 'the human mind has an appetite for fruit of every kind, physical as well as moral, and the real utility of Science is to afford gratification to this appetite'.[10] In all the frenetic efforts sometimes made to 'explain' scientific commitment in terms of economic greed or social aspiration this is a simple point which historians neglect at their peril.

An alternative interpretation is that the Geological Society really was, in the first place, primarily concerned with the practical benefits to be brought by a co-ordinated exercise in geological teamwork, and that it later became 'polite' as hangers-on, intent on seeking public esteem, joined its ranks and as the formality of its structure was seen as congenial both by those who belonged to the Royal Society and by those unlikely to do so. That practical considerations ran in the minds of the early members is suggested by the establishment of network research in response to its *Geological Inquiries*, and by the avoidance of much fruitless theoretical controversy along neptunist/vulcanist lines.[11] Against that must be set the low proportion of 'practical' papers actually published. The view that technical considerations did loom largely in the minds of early members has been put by P. J. Weindling, stressing the importance of the lobby for a National Collection of Minerals, based on a national survey covering the United Kingdom. Many of the subscribers to this Collection had a reputation for social concern and philanthropy, though little taste for radical politics, and Weindling has suggested that they, and the early Geological Society, sought to institutionalise geology not because the earth was in decay but because they feared that civil society was.[12] If this were all that was meant by 'utility' the position is virtually that suggested by Porter. However, Weindling distinguishes between the mining interest, which would appeal to landowners, and the collecting interest, which would be espoused by merchants. If the latter is taken into account, economic utility may well have been more significant than Porter suggests. The three founders of the National Collection were well to the fore in their attempt to involve the East India Company, and one of them (Greville) had been a Lord of Trade.[13]

In the end the Collection foundered, as Weindling says, on the rocks of self-interest of the Royal Institution's proprietors.[14] They even denied access to the Founders of the Collection unless they were prepared to pay the one hundred and fifty guineas necessary to become a proprietor.[15]

Thus in the early days of the Geological Society it is clear that utilitarian considerations did play their part, both of the narrowly economic kind and in a more ideological sense. Equally, however, separation from the Royal Society may be seen in terms of individual ambition, tactless authoritarianism and the sheer volume of new geological knowledge with which the Royal Society could not reasonably be expected to cope. As Martin Rudwick observes, this had 'far-reaching consequences for the specialization – and fragmentation of science',[16] even though the Astronomical Society was to face similar difficulties in 1820.

For the geologists the Baconian fact-collecting tradition was weakened for some years by those members of the Geological

Society whose ambitions were chiefly social. However, in the provinces geology was pursued with a more single-minded enthusiasm in the few specialist societies that existed outside London, or as part of the wider Natural History Movement. Despite the problems of isolation, the difficulty of getting things published, the frequent adherence to archaic theories and an understandable parochialism of outlook,[17] innumerable facts were gathered together in the best Baconian tradition, much as the promoters of the London Society had hoped.

In the far west of England, the traditional mining of tin was, from about 1770, being supplemented by production of copper and other metals. In 1814 the Royal Geological Society of Cornwall sprang into being, with the declared intention of using geological knowledge to exploit the region's vast mineral wealth.[18] Although not wholly achieving its objectives (e.g. to establish a local mining records office) it published *Transactions* which were a highly successful blend of pure and applied mineralogy. It is constructive to compare the progress of this Society with that of the Newcastle Lit. and Phil. which was set up at least partly to achieve similar ends. Although intended to provide useful scientific knowledge for mining engineers it fell far short of this objective, which is surprising since Newcastle, like Penzance, was in the heart of mining country. Both were remote from London, each had strong provincial consciousness. In both areas mine owners were often gentlemen of leisure and cultural leaders in the neighbourhood. Penzance, however, had two significant advantages. First was that several of its prominent mineralogists had had experience of the great mining institutions of Central Europe, especially John Hawkins and Henry Vivian. By contrast, many owners of coal mines in the north-east were sublimely ignorant of geology, and seemed content to remain so. A second advantage was that the Cornish institution was a specialist one, its Northumbrian counterpart being a generalist Lit. and Phil. The reasons for this are plain. Twenty-one years had elapsed since the birth of the Newcastle Lit. and Phil. when the Royal Geological Society of Cornwall was established, and during those years geology had in fact become much more specialised. Moreover, science as such was more obviously applicable to Cornish metal-mining than to coal-mining in the north east. Both metallurgical and mineralogical analyses are more obviously important in the former. In Cornwall, the young Humphry Davy and his friends had experienced the intoxicating pleasure of scientific discovery as they combined mineral prospecting with explorations of the Cornish coastline. If the Northumbrian pit-heaps generated similar enthusiasm the evidence is concealed from us. In other words the Cornish success may well have been dependent on the nature of the science being pursued, its degree of specialisation, its apparent utility and its aesthetic appeal.

Twenty-five years had to elapse before the formation of the next major provincial society to specialise in geology. This was the Manchester Geological Society. Although not far from the Lancashire coalfields and the lead mines to the north east, Manchester was hardly in an area of comparable mining importance to those of Newcastle and Penzance. Its early activists included many of the local 'devotees' of science, as William Fairbairn, J. E. Bowman and James Hayward. The last named was later to become President of the London Geographical Society. In a city where the Lit. and Phil. was (exceptionally) a powerful voice for science, it was all the more remarkable that a specialist Geological Society should be felt necessary. The reason appears not to lie in local rivalries. The two societies had overlapping membership and the Secretary of the Geological Society (E. Binney) was to become President of the Lit. and Phil. It seems that, once again, increasing specialism was responsible, a trend to which Mancunians seemed to be particularly sensitive.

From provincial geological societies we turn, finally, to other institutional expressions of the growing specialism in earth science. In the 1820s and 1830s the two ancient English universities gave geology official recognition with professorships at each, though students were not compelled to attend their lectures. Chief among English geologists of the 1820s was William Buckland, Professor at Oxford since 1819. An indefatigable upholder of catastrophist geology, he became famous as an advocate of 'diluvialism' – a belief that the flood described in Genesis was universal and responsible for geological features conspicuous today. His colleague at

Cambridge, Adam Sedgwick, was appointed to his Chair the previous year. Though he knew nothing of the subject when he started, he rapidly became proficient, and by 1831 had begun his major work of elucidating the Cambrian series of rocks. Both he and Buckland were clergymen and sought to reconcile geological data with Biblical texts, their 'Mosaic science' being some consolation to those who feared a confrontation between science and religion. Something like a confrontation came with the work of Buckland's most distinguished pupil, Charles Lyell, later to be Professor of Geology at King's College, London, and the propounder of a 'uniformitarianism' which implied a far vaster time-scale for earth history than the traditionally assigned six thousand years or so. Since the Scottish geologist James Hutton had proposed that in earth history 'we find no vestige of a beginning – no prospect of an end'[19] a certain tension had existed between theological and scientific views on this subject, although it would be wildly exaggerated to claim it as a conflict. But not even the efforts of a Buckland or a Sedgwick could completely dispel a certain aura of irreligion about geology, and this may go some way to explain its slower development than other branches of Natural History.

In 1815 William Smith completed his Herculean labours with his 'Geological Map of England'. Two years later a young man from Devon, Henry T. de la Beche, joined the Ordnance Survey and soon felt the need for updating Smith's work. On his own account he began a fresh geological survey of Britain. At first he was able to finance himself, but as the project grew he sought for government support and, with the backing of Sedgwick and Buckland, obtained Treasury grants for a Geological Survey of Great Britain and Ireland, though he had to supplement this considerably out of his own pocket. This was at first part of the Ordnance Survey, but it was 'demilitarised' by Sir Robert Peel who transferred it to the Department of Woods and Forests. One of its junior workers was A. C. Ramsay, a Scot who eventually became Director General.[20]

At a British Association meeting in Newcastle in 1838, de la Beche induced his friend Thomas Sopwith to urge the importance of National Mining Records. This was all the more important in view of the landowners' ignorance of geology, and the belief in such traditional aphorisms as 'no coal under limestone', which Smith had shown to be quite wrong. As a result of this appeal a committee was set up consisting of Sopwith, de la Beche, Buckland, Lyell, Sir Charles Lemon (a Cornish landowner) and the Marquess of Northampton. Their efforts were successful, and the Commissioner of Woods and Forests authorised the establishment of a Museum of Economic Geology near Charing Cross in London. In addition to the museum itself, based on de la Beche's private collection and opened to the public in 1840, there was a Mining Record Office which had been the primary objective of the British Association Appeal. In charge of this was T. B. Jordan, Secretary to the Cornwall Polytechnic Society. Finally an analytical laboratory was established with Richard Phillips as superintendent.

Expansion of all sides of this work led to a demand for new accommodation, and in 1851 the new Museum of Practical Geology was opened in Jermyn Street, Piccadilly, by the Prince Consort. Six months later de la Beche's final geological institution received royal sanction with the opening of the Government School of Mines and Science. This eventually became the Royal School of Mines, a direct ancestor of Imperial College, London.[21] Its creation must be seen as a result of many converging interests: the Cornish lobby, the powerful influence of de la Beche himself, the rising interest in science stimulated by the Great Exhibition of 1851, the personal involvement of the Prince Consort, and above all the independent recognition that matters were too urgent for the traditional self-help approach. A new specialist training was necessary. The kind of education required was new to Britain, though not to France. And so with Chairs in Chemistry, Natural History, Mechanical Science, Metallurgy, Geology and Mining and Mineralogy, the new Institution set its academic seal on a process of specialisation that fifty years previously would have been unthinkable.

11.3 CHEMISTRY FOR CHEMISTS

The emergence of geology as a specialised scientific enterprise, with its own institutions and ethos, came about despite the reluctance of those who might have gained the most from it: the landed gentry and mine owners of the north-east and elsewhere. By contrast, its sister subject, chemistry, became firmly identified as a specialism worthy of institutional recognition by people who had an obvious and direct interest in its success. And even here the crude economic model does not apply, for very different kinds of success were envisaged. Only in a few cases would the function of specifically chemical institutions have been seen as likely to promote immediate, large-scale economic gain. Equally, it does not seem that in any but the most general sense could they have been conceived to maintain the status of traditional groups such as landowners. The picture that emerges is rather one of several disparate groups perceiving a common interest in the advancement of specialist chemical research. By this means they would be able to gain a new sense of corporate identity, to establish their position in the international community of science, and gain self-confidence in developing the new and complex skills required in the chemical specialisms that were then emerging.

It is worth looking briefly, as G. K. Roberts has done,[22] at what she called 'the early-Victorian chemical constituency' in England. The reorganisation of the medical profession in the early years of the century, associated particularly with the Apothecaries' Act of 1815, helped to raise the standard of medical education, and that included training in practical chemistry. At University College, London, a Medical Faculty was established in 1826, and within three years was offering chemical training. The links with Scotland were strong, its syllabus being modelled on that at Edinburgh, and its first two chemistry professors (Edward Turner and Thomas Graham) having taught in Edinburgh and Glasgow respectively. By 1840 practical chemistry was part of the teaching programme for medical students at King's College, London, at four other London medical schools and at three in the provinces (Birmingham, Bristol and York).[23]

For at least two centuries what might be somewhat dubiously called the medical profession had been plagued by power-struggles, with physicians, surgeons and apothecaries manoeuvring for control or independence. The chemical task of making up medicines in the early nineteenth century was in the hands of an assortment of practitioners of widely diverse skills and abilities. A Bill introduced in 1841 to limit medical practice to those properly certificated included also the chemist and druggist. Although socially desirable in many ways, the measure was seen as an attempt to bring pharmacy completely under the control of the medical profession. The pharmacists, who were not inexperienced in the art of public protest and parliamentary lobbying, took up the challenge with such vigour that the Bill was talked out of Parliament, although the threat remained of a medical takeover by some other means.

Accordingly, in April 1841, the Pharmaceutical Society of Great Britain was founded under the leadership of a London pharmacist, Jacob Bell, and with the express intention of protecting the interests of chemists and druggists. Though without doubt a professional pressure group, the Pharmaceutical Society sought also to promote research and training in pharmacy's specialist techniques. It opened its School in 1842 and started to teach practical pharmaceutical chemistry in 1844.

Certainly there was never any question of the Pharmaceutical Society being a gentlemanly, polite institution. Neither did it owe much to popular pressures, from the thousands of one-man businesses that might have been expected to unite around its banner. In fact the chief impetus came from the large pharmaceutical wholesalers in London, firms such as Allen, Hanburys and Barry. Partly this reflects a greater metropolitan awareness of the threat from medicine, and partly the strong Quaker business ethic, for many of the promoters were wealthy members of the Society of Friends.

Another element in the chemical constituency was agricultural chemistry, where determined efforts were being made to develop a scientific basis for farming. Chemistry was prominent in these efforts.[24] As early as 1807 the Bath and West of England Society determined to establish its own Chemical Laboratory, and in 1843 the Agricultural Chemistry

Association was set up in Edinburgh under the auspices of the Highland Society; its chemist was J. F. W. Johnston of Durham University. In the same year the Royal Agricultural Society established an unsalaried post of Analyst and several private individuals began raising support for a teaching institution to include agricultural chemistry. As a result the Royal Agricultural College of Cirencester opened in 1845. Meanwhile, at a British Association meeting in 1844 a national project was launched to provide analyses of plant ashes. A 'new feature' of the Chemical Section of that meeting was the presence of several country gentlemen, whose scientific interest may have been not unconnected with the imminent prospect of the repeal of the Corn Laws and consequent need for increasing the efficiency of British farming.[25] But the chief impetus lay in the initiatives taken by the professional chemists, notably Liebig and Daubeny. As subsequent events proclaimed, the land-owning interest in agricultural chemistry took a long time to develop. As with the coal-owners, financial self-interest was not a determining factor in the pursuit of science, but rather the rationale for the scientists' public relations exercises. Indeed landowners who did consult a chemist were less likely to be interested in soils than in mineral resources.[26]

Mention must also be made of the variety of manufacturers whose activities were transforming the British landscape and social structure, and who were beginning to employ specialist chemists. In calico-printing the first textile chemist in a Lancashire print works appeared in about 1817, but for many years it was the owners rather than the employees who were 'gentlemen perfectly understanding chemistry'.[27] A few of the early Victorian gasworks had chemists on the staff (such as Frederick Accum in London and John Leigh in Manchester). Soap manufacturers needed chemical knowledge, and one or two had tiny laboratories. In brewing, Robert Warington was appointed chemist to a London firm in 1831, and Henry Böttinger to Allsopp's at Burton-on-Trent in 1843. But the total number of these practitioners was very small before the 1840s.[28]

Manufacturers of chemicals differed in their perceptions about the need for specialist chemical knowledge. Some, such as Christian Allhusen, who started his alkali works on Tyneside in 1834, became millionaires on the basis of almost zero training in chemistry; while others such as William Losh, Lord Dundonald, Peter Spence, William Gossage and Hugh Lee Pattinson were all successful proprietors whose place of chemical training varied from the University of Cambridge (Losh) to the local gas works (Spence). A works chemical laboratory was a rarity in the 1840s, but several very significant discoveries made under appallingly inadequate conditions were beginning to call attention to the deficiency. One may instance the absorption towers designed by Gossage (1836) and Glover (1859), spectacular examples of applied chemistry which led to a significant decrease in atmospheric pollution, and to a large increase in profits in both cases.

To some extent the formation of the Chemical Society of London was a response to pressure from 'professionals' of these kinds, though that is hardly half the story. Formally established on 30 March 1841 (and still surviving as the Royal Society of Chemistry) it became the first national society for chemical science anywhere in the world, being later followed by France (1855), Germany (1866), Russia (1869) and the USA (1876). Each of these institutions had a strong academic emphasis and concentrated on the dissemination of research results. The London Chemical Society (unlike its unsuccessful namesake of the 1820s) was no exception, and in the early days academics were about equally balanced by those who came from the non-academic worlds of industry, medicine, etc. Ironically for a society founded in London, the Scottish tradition was more influential than that of the metropolitan chemists.[29] Of the seventy-seven original members, Scotland (with fourteen) claimed the largest share after London (with forty).[30] But in the first Council, Scottish influence was proportionally stronger still, and was of course specially mediated by the President, Thomas Graham.

Academic chemistry in England, indeed in Britain, was at a low ebb. Given the absence from the *Philosophical Transactions of the Royal Society* in 1841 of a single chemical paper, one might well wonder why a Chemical Society should be needed. Even though the reputation of Thomas Thomson in Glasgow as a 'chemist breeder' might be obscured by that of Liebig, his was by far the most active research school in British chemistry.[31] As two writers complained in that year 'little of what is done abroad, especially in Germany, seems to find its way into England, or at least until the lapse of some years'.[32] The visit of Liebig to the British Association in 1837 may have been a critical factor in opening English eyes to the deficiency;[33] he was certainly the first foreign member to be elected to the Chemical Society. The Scots, unlike the English, had close links with continental chemistry and an awareness of the spectacular progress then being made in Germany, France and even Sweden. Clear evidence for the Scottish conception of chemistry can be seen in the exclusion of electricity from consideration – a further narrowing-down of what was understood by chemical specialism that was not at all in the spirit of Davy or Faraday (who joined only in 1842 and published little in the *Journal*). It was deplored by one potential member approached in 1841, the photographic pioneer H. Fox Talbot,[34] who thought that chemistry without electricity might not be sufficient to keep them occupied.

There were of course other local circumstances favouring the formation of the Society; the regrouping of pharmacists at the same time to form the Pharmaceutical Society could be deemed to have been a threat to the aspirations of the other kind of 'chemists'; the new penny postage made long distance communication much easier; and Robert Warington, the founder of the Society, had time on his hands between two official appointments, a need to use his considerable creative energies, and (along with seven or eight other founding members) experience of membership of the Spitalfields Mathematical Society.[35]

The Chemical Society may be seen as the embodiment of a Scottish academic tradition in a characteristically English institution, for, as we have seen, the metropolis had a multitude of scientific societies earlier in the century catering mainly for an amateur clientele. It was this paradoxical combination of specialism and amateurism that was so distinctively English and may be a sufficient explanation for the priority of Britain in establishing a national Chemical Society.

Four years later the Royal College of Chemistry marked a

further stage of specialisation in London, undertaking something the Chemical Society could not, or would not, contemplate: the education and training of chemists. Originally conceived by two little-known practitioners of science (John Lloyd Bullock, a chemist and druggist, and John Gardner, an apothecary) it started as a privately funded organisation, though with support from the Prince Consort. Through his influence it was to acquire its first Professor of Chemistry, the prestigious German organic chemist A. W. Hofmann. G. K. Roberts has identified the interests of about 560 of its 760 original backers. Of these the four biggest groups are landowners (27.3 per cent), chemists and druggists (19.6 per cent), medical men (17.7 per cent), and manufacturers (12.7 per cent). The first and last classes could clearly expect some kind of financial gain, as soil analysis and quality control respectively were becoming important to their enterprises. Whether they all did think of the College in these terms, however, it is not quite certain. But some landowners certainly seemed to expect that the College would be able not only to train chemists but also, in the short term, to act as an analytical consultancy for them. Of the medical people about two-thirds had Scottish qualifications and therefore an awareness of the relevance of chemistry to their profession. The chemists and druggists came mainly from the provinces and may well have represented a disenchantment with the metropolitan character of the Pharmaceutical Society.[36]

The Royal College of Chemistry, situated in Jermyn Street, opened its doors a year after the Pharmaceutical Society established its teaching laboratory and a year before University College and King's College (London) established professorships and opened new laboratories for teaching practical chemistry. All these were part of the new drive for specialist chemical training, and the promotion and growth of the Royal College of Chemistry may well have actively stimulated the other developments. In 1853 the College was taken over by the government and combined with the Museum of Economic Geology to form the Metropolitan School of Science, an ancestor of Imperial College.

The ultimate in chemical specialisation occurs with the rise of provincial societies devoted solely to chemistry. The situation in Newcastle-upon-Tyne is particularly instructive, not merely because the first of such institutions appeared in that city, but because two local Chemical Societies existed side by side. This fact, together with an analysis of membership trends, illuminates the changing social patterns of chemical practice.

Newcastle in the 1860s was one of the two most important chemical manufacturing centres in England (the other being Merseyside). The abundance of salt, coal and limestone, together with deep water access by the Tyne, helped to assure its prosperity. On its banks were a couple of dozen chemical works which supplied half the nation's alkali and bleaching powder. Iron and steel, made in the locality, were employed in the heavy engineering works of Charles Palmer, Sir William Armstrong and others. Steam engines, hydraulic machinery, armaments and of course ships were the more spectacular products of Tyneside industry. And in these industries most leaders recognised the need for chemical research and control. So, in June 1868, to cater for 'professional chemists and practical men' employed in local industry the Newcastle Chemical Society was established.[37] The initiative was taken by A. F. Marreco (Reader in Chemistry at the Newcastle College of Medicine) and R. C. Clapham (alkali Manufacturer and Quaker industrialist). Marreco was the son of A. J. F. Marreco of the ill-fated London Chemical Society. They appeared to share a passion for forming chemical societies, the son being conspicuously more successful. Over forty members were recruited, numbers doubled within a few months, regular meetings were established and *Transactions* were launched. At first discussions centred upon the dominant chemical process in the district, the Leblanc Process for making soda. After a few years they widened to include other topics, mostly of industrial importance; several important new processes were described or demonstrated and it was this Society which witnessed the first public demonstration of the electric light bulb, invented by J. W. Swan.

Put in these terms it sounds as though the Newcastle Chemical Society was a living embodiment of utilitarian values, with science applied to a specialised industry to the great benefit of both. That this was intended there can be no doubt, but several factors conspired to frustrate it in some measure. One of these, it must be said, was the class structure of the Society. Behind its formation lay the powerful support of wealthy industrialists like I. L. Bell, Armstrong, and many others. Even more important, the Society was clearly an unofficial offshoot of the Newcastle Literary and Philosophical Society. Of the twelve members of the first Committee, all except two were also in the Lit. and Phil.; one of these was to join later and the other was in the related Natural History Society. Almost exactly half the original membership were also in the Lit. and Phil. This decided the framework within which science was discussed, that of a traditional learned academic society, perhaps not appropriate to the actual problems of the chemical industry. These, if they did not involve fundamental research into alternative techniques, were more often than not of a trouble-shooting kind of a quite pedestrian nature. Yet the approach was entirely characteristic of the Lit. and Phil. and its other offshoot societies.[38] Most of the papers were from the inner circle of committee members, but by 1873 the President (George Lunge) was lamenting the lack of contributions.[39] There is little evidence that any people outside the Lit. and Phil. network would have contributed. The Vice-President, John Glover, felt constrained to deplore the general state of the English manufacturing industry, as indicated by its bad showing at the Vienna Exhibition. He attributed this partly to the fact that science had not yet acquired its proper place in the minds of manufacturers. But he also blamed rising coal prices and 'the lower moral tone of our national life'.[40] Though no capitalist, he seems to have been referring to industrial unrest.[41] That was how the President understood his words, admitting that men ought to be better paid but deploring paternalism. He regretted the loss of 'command upon the live tools that were requisite to carry out manufacturing operations'. This evil 'could be counteracted to a very great extent by very close scientific supervision'.[42] This, then, was part of the Society's strategy: not 'social control' in a nebulous, general way, but a much more specific control of an unruly work force brought about by reducing the latitude allowed them by traditional pragmatic practices:

> When the rule of thumb fails . . . science must step in to supplement it.[43]

Too much can be made of this, but it does suggest that the Society was identified with the owners and managers rather than the foreman and labourers, and membership data support this.

This view is borne out by the existence from 1870 of a rival organisation, the Tyne Social Chemical Society:

> Formed at first merely as a social gathering where a few chemists and managers of adjoining works met in the evenings at each others houses to talk over in a friendly way the difficulties and mishaps met with in the course of everyday work, and to suggest from each others experience the proper methods of surmounting them; it was soon found necessary to enlarge the circle of members and to meet in a fixed central place, in order to suit the convenience of members who in many cases reside in rather out of the way places near their manufactories.[44]

The word 'social' was eventually dropped from its title, but it continued to be mainly a self-help group for the junior managers and foremen concerned with the day-to-day running of chemical plant. It was much smaller than its rival, their maximum numbers being respectively 40 and 155. The Lit. and Phil. connection was thin (perhaps six members) and about eight belonged to both chemical societies (including Lunge). Overtures were made by the Newcastle Chemical Society in 1876 'relative to a proposed union',[45] but without success. Four years later, however, amalgamation did take place. By that time the character of the older society had markedly changed. Whereas at the beginning the ratio of employers and landowners to employees had been about 3:2, by 1882 that figure was 2:3. Also, many employees at first had been fairly senior managers, but it seems that junior management predominated at the end.[46] In fact 1882 was the end of the Society, for it was to become the Newcastle Section of the newly formed Society of Chemical Industry. This in its turn sprang from a local Chemical Society in South

Lancashire which was deliberately modelled on the pattern in Newcastle.[47] Like the Chemical Society this was (and is) a national body, but (like those pioneer societies in Newcastle) its constituency is overwhelmingly industrial. Specialism had gone one stage further in response to the perceived needs of an industry feeling strongly the cold winds of competition from a German competitor that was better organised and better informed.

While no institution quite as specialised as the Royal College of Chemistry appeared in the north, a College of Physical Science was established in Newcastle in 1871, with strong support from the Newcastle Chemical Society and the Lit. and Phil. Amongst its teachers was the new Professor of Chemistry, A. Freire Marreco. The College went through various metamorphoses until in 1964 it became the University of Newcastle-upon-Tyne.[48]

11.4 SPECIALISTS OVERSEAS

Even a brief comparison between French and British attitudes to scientific specialism reveals two major differences of institutional practice. The first is that, generally speaking, French specialist scientific societies were founded considerably later than their British counterparts. The following table shows a dozen of the leading national representatives. Only three out of the twelve societies preceded their English counterparts. The gap between the institution of English and French societies varies from 67 to minus 9 years.

From this it is quite improper to draw conclusions about the relative inferiority of French science which, in some respects at least, was ahead of its British counterparts. Instead one needs to see these phenomena in the light of the wider social structure of French science in the middle of the nineteenth century.[49]

Robert Fox[50] has pointed to the relatively minor role played by French national societies as compared with that of the great central institutions as the Académie or Muséum. In England the Royal Society had not yet acquired its present international standing, both as a forum for scientific discussion and as a publishing body. Even its celebrated *Philosophical Transactions*

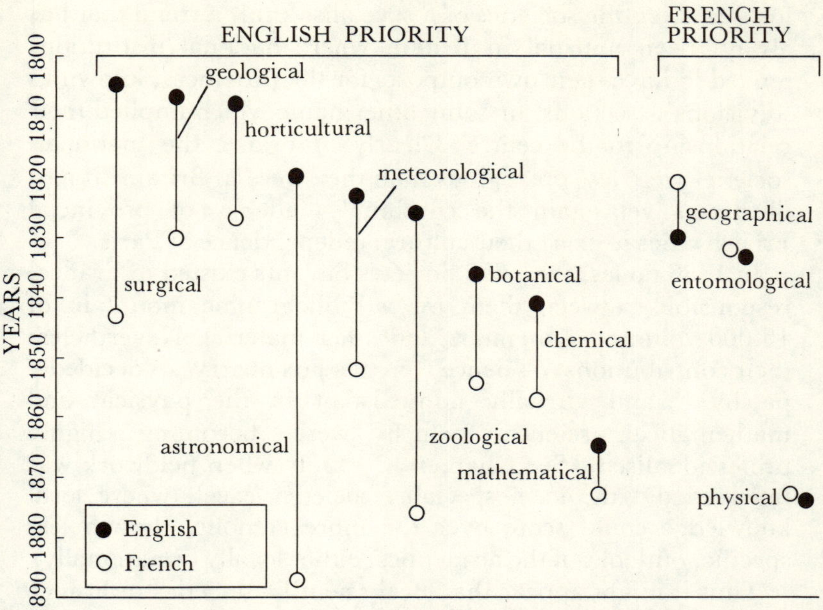

FIGURE 6 Dates of Formation of National Scientific Societies.

catered for only a fraction of the specialist science being published in Britain in the middle of the century. In France the first generation of national societies dealt mainly with those serving the needs of an amateur membership, while the second generation (for mathematics, physics and chemistry) tended to be more 'professional' and were soon publishing prestige research. Nevertheless even they were constrained by the authority (and the subsidised periodicals) of the traditional institutions, to which it must be said they posed little serious threat. Fox also contrasts the early struggles of the Société Zoologique with its opulent analogue across the Channel, ten times larger and enjoying the revenue from its zoological gardens in Regent's Park. Yet that cannot alone explain the relative fortunes of the two institutions, nor can it account for the greater success of the British Chemical Society as compared with that in France. One is compelled to fall back on more general and fundamental explanations.

A second difference is the much greater French predilection

for *local* scientific societies of a specialist kind, a trend that has always been unusual in Britain where national institutions tended to have their own outposts for the provinces, known as 'divisions', 'sections' or some other name which implied their relationship to the centre. Clearly in France the 'national' societies were less prestigious than they were in Britain. It also illustrates yet again the continual tendency of provincial French cities to exert their cultural independence of Paris.

In 1886 no less than 655 Sociétés Savants existed in France, responsible between them for a bibliographic mountain of 15,000 volumes of memoirs and other material. Nevertheless their contributions to science were, as Fox observes, 'decidedly patchy',[51] and virtually non-existent in the physical and mathematical sciences which were becoming highly professionalised. (See Chapter 12.) Only when fieldwork was involved did the local specialist societies excel, where local knowledge could score over the more catholic, though less specific, outlooks of the academics, either locally or nationally.

Thus it might appear that local scientific societies in France served a similar purpose to the natural history societies and the Literary and Philosophical societies of England several decades earlier. Scientific research, except in the limited sense of data collection in the field, was not notably promoted, though scientific awareness must have been enhanced by the avalanche of published reports. To that extent the English parallel is a valid one. But in other respects was this merely a continental version of the English use of science as 'ornamental learning' undertaken in the pursuit of such non-scientific ends as fashion and personal status? Caution is needed in pressing the parallel too far. In the first place, as we have seen (Chapters 6 and 10), it is highly doubtful whether much English provincial science in the first half of the nineteenth century *can* be adequately understood merely – or mainly – in these terms, fashionable though it may have been to do so some years ago. Second, it is vital to distinguish between intention and achievement. Merely because the 'hard' scientific specialisms did not prosper in the local *société savant* it is not to imply that prosperity was never intended. Third, as Robert Fox has again reminded us, the proliferation of these societies is not only one of the most startling cultural phenomena of nineteenth-century

France; it is also one of the most neglected.[52] Much more detailed analysis is necessary before one can speak with confidence of a British parallel. The very fact that so many of them existed speaks for a large measure of caution. A fourth objection, and one of considerable force, is that, unlike the académies, the new specialist societies almost always had an open membership. That being so it is hard to see how membership could be regarded as a symbol of status, or even a refuge from social 'marginality'. If this view of their social function is of doubtful value for Britain, it is surely less useful still for France. Finally, a narrow 'cultural' interpretation is contradicted by a declared dedication to technical objectives, the contents of published *Bulletins* and a rich and research-orientated programme. Such is the case with the Société Industrielle of Mulhouse, founded in 1826 and throughout the century unashamedly committed to the chemistry, physics and mathematics that were needed for the technical development of textile and other interests in the locality.

So strong was the same emphasis at Lille that the President of its Société des Sciences felt constrained publicly to warn his members against associating science too closely with utility.[53] Indeed, such an admission is itself testimony to a wider conception of the uses of science. But that it could ever have been deemed necessary suggests that it was, in fact, precisely the technical aspects that were, in this case at least, obtrusively to the fore. Finally, the very specialism of some of these provincial societies militates against the view that science was simply pursued as a means of enhancing the social status of its practitioners. This could be done a great deal more easily in the more generalised institutions and still more in those devoted to literary and artistic pursuits. The most prolific decade, the 1870s, saw at least 143 new societies launched in the French provinces. An analysis by Fox[54] gives the following figures for those which had anything to do with science (but excluding archaeology, geography and psychology):

Multidisciplinary (including humanities also)	22
General science	17
Natural History	3
Physical Sciences	2

Meteorology	2
Medicine	6
Pharmacy	5
Veterinary Medicine	1
Technology	7

Even taking into account agricultural and horticultural societies (not included in this analysis) the number of new societies devoted to science alone was less than one-third of the total provincial learned societies founded in that decade. But over half of these were specialist societies and that is still a significant number, suggesting a need that was heavily dependent upon specific content and methodology. A similar position was evident at Lyon, in 1886, when twenty-six local societies existed of which eleven dealt with pure or applied science.

The emergence of specialist scientific societies was not, of course, uniquely a French or British affair. In Germany and the United States, and elsewhere, discipline-based societies were being formed to serve the increasing band of professional or semi-professional workers who needed them for their job rather than for their leisure. To conclude this Section the national societies for chemistry in Paris,[55] Berlin[56] and New York[57] are briefly discussed.

One obvious point about their formation is that there was in every case a response to a very local, not to say parochial, demand. In no instance were grandiloquent plans for national status sufficiently well advanced to obscure more immediate and local needs. However, the extent to which this is true varied with the different societies. It was particularly true with the Société Chimique de Paris, founded in May 1857 by three young chemists as a self-help group to promote their chemical studies. One of them (Arnaudon) was a 'préparateur' to M. Chevreul, doyen of French chemists, in his industrial laboratory at the Gobelins; another (Collinet) performed a similar function for the eminent J. B. Dumas in the Paris Faculty of Sciences; a third, (Ubaldini), worked in the laboratory of the Collège de France. Moreover, the first two Presidents were, as the official centenary volume records in some embarrassment, only distinguished 'préparateurs'.[58]

They were succeeded by Aimé Girard, 28 years old and chiefly interested in industrial chemistry. The British analogy which most rapidly springs to mind is not the famous Chemical Society in London of 1841, but the obscure London Chemical Society of 1824, where initiative diffused from the grass-roots upwards.

Ten years later the prospects in Berlin for a major new development were a little more auspicious. About one hundred chemists who met at the Museum of Industry on 11 November 1867 were told by the organic chemist Adolf Baeyer that the moment was 'most favourable for the union of Berlin chemists, since for the first time in its existence the University has obtained a chemical teaching laboratory'.[59] This innovation, which was obtained at the instigation of the newly arrived Professor of Chemistry, A. W. Hofmann, was a provision which his predecessor Mitscherlich, had been quite unable to exact from the Prussian Government. It was significant because of the burgeoning chemical dyestuffs industry with its wholly unprecedented demands for new chemicals and for skilled labour to produce them. Other centres had teaching laboratories, most notably the one at Giessen founded by Liebig in 1825. But they also had their associations of chemists. The foundation of the Berlin Chemical Society must therefore be seen as a chance to catch up with German rivals, and a provision for a much needed research tool for the newly trained chemistry students waiting to serve in industry.

An American counterpart to these European institutions can trace its origins to a different kind of local circumstance. To celebrate the centenary of Priestley's discovery of oxygen, seventy-five chemists from many American states gathered at the English chemist's last home in Northumberland, Pennsylvania, on 31 July 1874. Such was the intellectual and emotional warmth generated by this anniversary that a proposal for a national Chemical Society was mooted, though not carried (partly on the grounds that chemical interests were already served by the chemical section of the American Association for the Advancement of Science). However, eighteen months later, in January 1872, several of the same chemists met at the home of C. F. Chandler, Professor of Chemistry at the Columbia School of Mines, New York, and initiated moves to create a local chemical society in the area of New York.[60]

In all these cases it is not hard to discern the awareness of the precedent created by the Chemical Society of London, nor of a conscious attempt to take it as a model. In France the youthful self-help group was not long to remain as such. According to Gautier transformation was gradual as 'little by little' the up and coming great ones of French chemistry associated themselves with the Society.[61] Something like an academic takeover seems to have occurred rather more quickly, for by 1859 the President was J. B. Dumas himself; the Vice-Presidents included Pasteur, Thenard, and Berthelot. Even before this one of the Secretaries, A. Wurtz, had established a *Bulletin* and another research journal. By this means, and by enlisting the help of his academic allies, Wurtz had, as Gautier claimed, 'transformed a circle of students under instruction to a veritable learned society'.[62] And in this transformation there is no doubt that Wurtz, a Foreign Member of the London Chemical Society since 1856, was consciously taking that Society as a model. So Gautier categorically asserted in 1891, adding that Wurtz came to London for the very purpose of gathering information to assist him in that task.[63]

The German debt to London is even clearer. The moving spirit, Hofmann, had been a President of the Chemical Society in 1861 and several of his pupils had come from London with him. And he reminded the inaugural meeting that the London Society had not only the interests of its members at heart, but the progress of science generally. Berlin would do well to follow suit. In New York the initial attempts to raise interest for a local chemical society were so successful that Chandler and his friends wrote a further letter to 220 chemists nationwide seeking to extend the concept to a *national* society, and citing the three precedents already in existence. As a result the American Chemical Society was formed on the 6 April 1876. Its first President, J. W. Draper (famous for his photo-chemistry and notorious for his historiography)[64] was an English emigré. At least one Vice-President (J. W. Mallet) owned British citizenship, and the Treasurer, W. M. Habirshaw (an analytical chemist), was not only a Fellow of the Chemical Society but also belonged to its provincial neighbour in Newcastle. Chandler (who published the *American Chemist*) had links with the *Chemical News* of England. What is surprising is not that the Americans followed the British example but that they took so long to do it. The new sense of nationhood following the end of the Civil War undoubtedly had something to do with that.

Finally, the three societies eventually became national institutions, though in different ways. The American Chemical Society was so called at the beginning, though its early history was dogged with conflict between 'resident' (New York) members and others, who resented the dominance of New Yorkers on committees. That, together with immense problems of communications in a vast country, led to the establishment of rival local chemical societies, notably at Washington in 1884. Misgivings were somewhat allayed by a new constitution of 1890 which instituted postal voting and local Sections.

In the German example, aspirations to national status were strongly encouraged from the first by Hofmann, who pointed out that Berlin was neither more nor less than one of many centres of intellectual life in Germany, so merely to speak of the Berlin Chemical Society would not guarantee any national recognition. London could stand for England, Paris could stand for France, but this was not so with Berlin.[65] In a sense he was right about France, with its strong metropolitanism. Yet his view was that of a foreigner, and would not have accorded with the views of French provincial academics. Perhaps that is why the Société Chimique de Paris remained as such until 1906, when it became Société Chimique de France, and then only at the request of its provincial contemporaries. But from the beginning Hofmann's brainchild had it both ways: Deutsche chemische Gesellschaft der Berlin.

12. The Road to Professionalisation

12.1 SCIENCE AS A PROFESSION

In 1799 Humphry Davy wrote 'philosophy, chemistry and medicine are my profession'.[1] Two years later he rejoiced that being paid to pursue science at the Royal Institution relieved him of 'the obligation of labouring at a profession'.[2] For him 'profession' meant something like 'paid employment', and little more. When, however, Alexander Findlay doubted 'whether one could speak of the existence of a profession of chemistry before the founding of the Institute of Chemistry in 1877',[3] he used the term in quite a different sense. This variety of usage has plagued English writing, for the term has considerably changed its meaning over the last 180 years. Today a professional is understood as a member of a community united by a common avocation. This membership implies several additional characteristics, including:

(a) An agreed minimum standard of intellectual competence.
(b) Acceptance of certain social obligations (like the Hippocratic oath for medicine).
(c) An expectation of a certain level of remuneration and a recognisable career structure.
(d) A degree of recognition by other professional groups and by government.
(e) A strong sense of corporate identity.

Thus in exchange for certain guarantees, (a) and (b), society is expected to recognise the claims of the professional, (c) and (d), and all combine to give a sense of identity within a well-defined community (e). In so far as any of these features may be seen in a vocational pursuit, that pursuit may be said to be on the way to professionalisation in this modern sense, for such changes are gradual rather than sudden. While it is quite unhistorical to imply that all these criteria were in the minds of earlier advocates of professional science, there is no doubt that some of them were, and that, before the nineteenth century closed, each of them had been canvassed in one way or another.[4]

When Babbage complained in 1830 that 'the pursuit of science does not, in England, constitute a distinct profession',[5] he implied rather more than Davy; he lamented the lack of recognised career structure and of proper encouragement by learned societies and by the state. Chemistry was a case in point, though by no means the only one.

In 1820 the Court of Common Pleas heard the case of Severn, King and Co. (sugar bakers) *versus* several insurance companies, concerning liability for losses in a disastrous fire at the firm's Whitechapel premises. Many chemical witnesses were called, including W. T. Brande, J. T. Cooper, F. C. Accum, T. Thomson and Michael Faraday. Since a witness had to have 'professional status' in order to claim expenses, this became a further matter of contention. In November 1821, the court ruled that the man of science, without recognised learning and actually basing arguments on the novel process of 'making experiments', was no more professional than a mechanic. In thus denying science equivalent status to law, medicine or the Church, the judges were reflecting the established social values of the ruling class (especially as many of the witnesses on this occasion lacked an Anglican commitment as well as an Oxbridge training). The decision settled for many years to come the legal-professional status of a chemist,[6] despite the fact that chemists frequently made large sums of money as expert witnesses in litigation.[7]

Whatever the law might say, the status of scientific workers remained problematic for many decades, and the confusion is reflected in the presentation of census data. Thus, while in

1841 Britain had 10,853 chemists and druggists, it also had 684 manufacturing chemists, and several thousand other persons who might or might not have been doing chemistry (bleachers, soap-makers, manufacturers of soda, dyes, copperas and so on). Similar confusion persisted in 1851, and in 1861 manufacturing chemists were placed in the 'industrial class', along with artisans, while 'chemists and druggists' were in the 'professional class' which 'includes those

TABLE Membership of British Scientific Societies, 1868.

Mathematical and Physical Societies	
Statistical Society	371
London Mathematical Society	111
Royal Astronomical Society	528
Chemical Society	192
British Meteorological Society	306
Geological Society	1100
Geologists' Association	230
Meteorological Society of Scotland	520
Manchester Statistical Society	162
	3520
Biological and Natural History Societies	
Ethnological Society	219
Anthropological Society	1031
Linnean Society	482
Entomological Society	208
Royal Horticultural Society	3595
[Royal] Zoological Society	2923
Royal Botanical Society	2422
Royal Agricultural Society	5525
Edinburgh Botanical Society	368
Glasgow Natural History Society	120
Yorkshire Agricultural Society	500
Ulster Chemical Agricultural Society	218
Wiltshire Natural History Society	313
	17924

persons who are rendering direct service to mankind, and satisfying their intellectual, moral and devotional wants'.[8] Also in this class was a small group of 'scientific persons' whose identity must remain totally obscure.

More illuminating than census returns is an analysis by the economist Leone Levi of membership of scientific societies in 1868. He concluded that only 0.15 per cent of the population was then contributing to the advancement of science – a mere 45,000 individuals.[9] The above table gives his figures for society membership in the 'pure' sciences only. As Cardwell remarks, most of what Levi classified as 'science' was 'not only an amateur activity, but, in many respects, a dilettantist one as well'.[10]

None of these societies attempted to take on any of the attributes of a modern professional institution, except perhaps to give a general sense of corporate solidarity. Yet outside pure science, attempts had been made by other groups to break the legal/medical/clerical monopoly. The fact that this was first done by engineers suggests that professionalisation may well be easier for those with strong public visibility. The Institution of Civil Engineers was founded in London in 1818, with Thomas Telford as first President. Preoccupation of its members with canals and roadbuilding led to opposition from the railway engineers, especially George Stephenson who became first President of the rival Institution of Mechanical Engineers in 1847. Other engineering institutions followed as:

Institution of Gas Engineers 1863
Iron and Steel Institute 1869
Institution of Telegraph Engineers 1871
Institution of Electrical Engineers 1871
Institute of Marine Engineers 1889
Institution of Mining Engineers 1889
Institution of Mining and Metallurgy 1892

Public visibility was not the only condition. A second, though related to it, was size. Membership figures for the Institution of Mechanical Engineers indicate how they hovered for eight years around the 200 mark, but from 1855 entered a period of steep and almost linear growth. This 'induction period' suggests that small institutions are likely to face a frustrating period of 'acclimatisation' before significant expansion can occur.

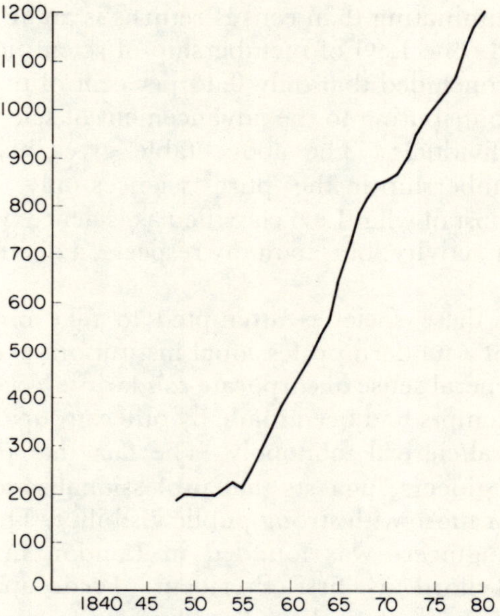

FIGURE 7 Membership of the Institution of Mechanical Engineers, 1840–1880 (all grades of membership are aggregated).

Thirdly, there must be a strong external pressure to unite, one that is therefore likely to affect more than one would-be professional group. Such a pressure can be generated by economic recession or simply by a band-wagon effect, with all trying to jump on board. The decade beginning in 1867 is an excellent example. That year and the next witnessed a major industrial recession, partly a legacy of various wars in Europe and partly a result of financial speculation at home. As the Editor of the *English Mechanic* put it 'morally and physically the year 1867 has had few equals since the era of the last great war and we fervently trust we will never see its like again'.[11] There was considerable industrial unrest (leading to the concession of a nine-hour day for engineering workers in 1871), and the formation of the Trades Union Congress in 1869. Industrialists had also been fairly shaken by the poor showing of British industry at the Paris Exhibition of 1867, revealing how far Britain was falling behind its European rivals.

And so there was a reaction. F. M. L. Thompson asserted:

> The very Victorians who condemned Trade Unions as vicious, restrictive, futile, and as unwarrantable interferences with individual liberty, flocked to join professional combinations.[12]

Thus in 1868 the Institution of Chartered Surveyors was formed. As its President said, everyone else having a common occupation had joined associations so why not the Surveyors?[13]

It is at this point that the stirrings began in the scientific community that were to lead to the first institution for the professionalisation of science. This was the Institute of Chemistry, eventually founded in 1877, not only the first, but also (in the nineteenth century) the only major organisation of its kind for science. Its formation demonstrated a response to all three stimuli referred to above.[14] Chemistry was now making a more visible impact on society than before; there were about 13,000 manufacturing chemists and many analysts and academics; and apart from the general unease of the late 1860s and early 1870s, a special kind of threat was being posed by the pharmaceutical profession. After passage of the Pharmacy Bill of 1868 a professional monopoly was granted to the pharmaceutical chemists, whose example was a powerful incentive to others who practised chemistry, but whose exclusive claim to the title 'chemist' helped to goad them into seeking a new professional unity themselves. And all this was happening against a background of animated debate on 'the endowment of research', and the dangers as well as the advantages of Government involvement with science. (See p. 240.) The scientific community was feeling extremely vulnerable.

12.2 THE FIRST INSTITUTE FOR PROFESSIONAL SCIENCE[15]

Since 1841 the chemists of Great Britain had enjoyed the distinction of having had the world's first national Chemical Society. It provided them with a forum for discussion, a *Journal* in which to publish their researches, a Library, a house in London but little else. Its tranquil life was rudely disturbed in 1867 by a series of events that raised, apparently for the first time, matters of great importance for that Society's future. It had been the practice in the Chemical Society, as with many other learned societies at that time and since, for the outgoing Council to nominate its successor, but with the proviso that every Fellow could, if he so desire, substitute a name of his own suggestion in place of one proposed by the Council. What took place at the Annual Meeting of March 1867 was strictly constitutional, but provoked a feeling of outrage that was well conveyed in the Editorial of 5 April in *Chemical News*. Words could scarcely be found to condemn the 'young gentleman of little more than a year's standing in the Society' who had the temerity 'to nominate another President, another Secretary, another Treasurer, and sundry other members of Council. . . . As it happened, however, the friends of the Society mustered in force, and, by a majority of nearly 4 to 1 decisively crushed the unwarrantable opposition.'[16] Events of this kind, repeated sporadically over the next few years, served to throw into strong relief the responsibilities and meaning of the Fellowship, as did elections to the Fellowship itself.

Matters came to a head in May 1876 when six out of nine proposed candidates for Fellowship were rejected. Clearly something was seriously amiss, and the Society's troubles were paraded before the readers of *Chemical News*. Writer after writer testified to the low standing of the Society. It was stated by one critic that the Chemical Society was 'not a body of chemists, but a motley mixture of all kinds of dabblers in science, and often, alas! very often, not even that'.[17] Such 'dabblers' would then wreak havoc in elections. Another enquired 'whether there was any bye-law which would give the Council some control over the use which Fellows might make of the privilege of membership', citing three or four flagrant instances in which the letters FCS had been used for purposes of advertisement.[18]

In fact there was no such legal restraint, and herein lay the Society's dilemma. As one member put it, chemistry was 'a profession as well as a science, and some members of the Chemical Society think that the Society should be the guardian of the professional as well as the scientific functions of chemistry'.[19]

Such professional awareness usually requires some immediate pressure to be transformed into action, and in the case of the chemists this was supplied by the sudden surge in demand for analysts. These were needed for monitoring levels of pollution in air and water, for ascertaining degrees of adulteration of food and medicines, and for quality control in industry.[20]

On 21 January 1876, *Chemical News* had carried a letter from C. R. A. Wright 'on the necessity for organisation amongst chemists, for the purpose of enhancing their professional status'. He observed that poorly qualified men would, by their incompetent analyses, bring discredit upon all. And with inadequate standards went inadequate recognition. The organisation movement that led to the formation of the Institute of Chemistry may be traced to that letter.[21] In the meantime chemists were becoming uncomfortably aware of further constraints upon them.

In February 1876 *Chemical News* carried a review of the 1875 *Year Book of Pharmacy*, and the anonymous reviewer concluded his remarks with the following:

> We think it is impossible to glance through this volume without being struck with the amount of influence and prestige which have accrued to the pharmaceutists of Britain from union. Will the 'analytical and consulting chemists' never follow this example to protect themselves against the encroachments which they suffer on various hands? We have often harped upon this string, but we wish to tune it.[22]

Since 1868 the pharmacists had been a shining example of professional unity – and professional monopoly. But the chemists were not so favoured, being subject to encroachments on at least two sides.

First, there were numerous cases where chemical analysis was being attempted by those whose training and experience were in the field of medicine. On occasions, as Wright observed, such work was undertaken by 'medical men possessing only the small modicum of chemical knowledge gained during their studies in the medical school, and often wholly innocent of any notion whatever how general quantitative work should be conducted, although they may have picked up some acquaintance with the methods to be adopted in the examination of some few articles of food, of water, or air, &c.'.[23]

A. H. Allen expressed strong opinions on 'the pseudo-scientific medical men' as a 'class generally very ready to quack as chemists, though remarkably tenacious of its own privileges'. He was also concerned with encroachments by unqualified druggists, 'Some years ago I heard a druggist say in the witness-box that he was an analytical chemist, though I have good reason to believe he never made an analysis in his life'.[24]

A *Chemical News* editorial identified another kind of threat.

> The Board of Trade have recently appointed a Water Examiner. Now a Water Examiner should be a chemist and a microscopist. Yet the gentleman appointed is an engineer! Turning to a closely connected matter, what can be more evident than that the treatment of sewage is a chemical question? The distinction between sewage and pure water is a chemical distinction. The methods of removing the impurities depend upon chemical principles, and chemistry alone can decide whether any sewage or refuse water is sufficiently purified to be safely admissable into rivers. Yet at a meeting of the Society of Engineers, recently held in Westminster Chambers, the President, Mr Vaughan Prendred, in the course of his address observed: 'The sewage question was the great question of the day, and it would require their (the engineers') deepest thought and utmost skill'. . . . The great principle of the division of labour, still more important in intellectual matters than in

> mere manual affairs, forbids any man from being at once chemist and engineer.[25]

The following issue carried a report on a further outrage:[26]

> The Government has employed Mr F. J. Bramwell, an engineer, to report on the possibility of the Mint proving a nuisance to the neighbourhood. Now, as the nuisance, if any should arise, must depend on the escape of fumes and gases, the question is strictly chemical, and can no more be solved by an engineer than by a lawyer or a clergyman.

For many years the engineers had looked upon chemists as a lesser breed of men. The advice given in the 1870s by W. Menelaus, President of the Iron and Steel Institute, was perhaps less extreme than it sounds:[27] chemists 'should be kept in a cage until something went wrong!'

Finally, the worst encroachment of all was from those unqualified in any field. Kingzett was exaggerating when he wrote of 'quacks who rejoice in the possession of bogus degrees'[28] as a serious threat to the livelihood of chemists, but there undoubtedly were cases where charlatans were using an authentic FCS as a token of excellence.

These controversies may seem reminiscent of the more fatuous 'demarcation disputes' of modern times, and they sprang from very similar causes: a deep sense of insecurity and a rising feeling of injustice. But in fairness one must add that the chemists had solid grounds for believing that such encroachments would genuinely lead to a debasement of analytical standards and real disservice to the public.

So in February 1877, the Institute of Chemistry was born,[29] with these objectives:

1. To promote and encourage a thorough study of chemistry and the allied branches of science in their application to the Arts, Manufactures, to Agriculture, and Public Health.
2. To ensure that persons practising as Consulting Chemists, or as Analytical Chemists, are qualified by study and training for the competent discharge of the duties they undertake.
3. To carry out measures necessary for the maintenance of the profession on a satisfactory basis.

The new body with its Associates and Fellows, was to be independent of the Chemical Society to which it owed its birth. Of its fifty-four founder members all but two were Fellows of the Chemical Society, and thirty-two had held office in that Society (four as President). After much heart searching and some legal advice the Chemical Society Council had decided that the Charter of the latter presented insuperable difficulties to some kind of select Institute within a larger Society. Just about half the membership were academics, but about ten were analysts and most of the others were in industry of various kinds; eight had medical connections and there were five pharmacists. The membership included seventeen Fellows of the Royal Society. Nearly seventy per cent of the founders lived in London. The first President was Edward Frankland.[30]

The subsequent history of the Institute need not detain us here. It is perhaps sufficient to say that it lacked legal authority to grant certificates of competence until the Royal Charter of 1885. Deliberately avoiding any suggestion of trade unionism, it lacked the capacity to negotiate salaries and its role was nearly always advisory. Its qualifying examinations, though they were well thought of, were never the only route to a career in chemistry. Its history has been dogged by conflict between academic and non-academic chemists, but it completed its centenary before uniting with the Chemical Society to form the present Royal Society of Chemistry (1980). Historically its importance is that it marked the first institutional recognition of the professionalisation of science in Britain. No comparable organisation came into being in the nineteenth century. The Institute of Physics was formed in 1920 and the Institute of Biology later still. It is because chemistry has both its academic and its applied aspects so clearly visible that its professionalisation took place at a time when the practical engineers had institutionalised their profession but other sciences were regarded as exclusively the province of the academic or the amateur.

12.3 POSTCRIPT: THROUGH A GLASS DARKLY

The relative merits of large and small telescopes may not seem a topic to produce deep social dissension, but in the closing years of the nineteenth century it turned out to be a touchstone in the social differentiation of British astronomers.[31] Until the establishment in 1881 of Norman Lockyer's research group at South Kensington, the merest handful of British astronomers had been professional. But in the subsequent years it was they who, in their Royal Observatories or University Departments, naturally had the larger instruments. In 1885 the English amateur astronomer W. F. Denning produced a book *Telescopic Work for Starlit Evenings* in which he gave numerous technical reasons for preferring small telescopes, particularly on the grounds of their optical advantages. One of these was the absence of 'glare' caused by an excess of light admitted by the larger lenses. Thus it was through such instruments of modest size – certainly with reflectors smaller than 12½ inches – that most contemporary work had been done on planetary markings. His arguments were polemically stated and incorporated dubious geometrical optics and they met with a chorus of dissent, particularly from those with larger instruments. Most of the critics were professionals, either in Britain or the USA.

As Lankford argues, the dispute was not just about the finer points of telescope construction or use, but also concerned the relative contributions to be made by amateur and professional astronomers. The Royal Astronomical Society included most of the latter, and from about 1875 its Gold Medal tended to go in their direction, rather than towards the dozens of gifted amateurs. In 1890 the British Astronomical Association was established for those who found the high subscriptions, academic standards or male exclusiveness of the Royal Astronomical Society not to their liking. This was not a splinter group, but it set out to offer 'a means of direction and organisation in the work of observation to amateur Astronomers'.[32] Nevertheless, by the 1890s, as Lankford observes, 'there were now two plainly marked roads. One led to professional status by way of higher education. . . . The other led to membership in the BAA.'[33] The competition was reduced as each accepted a distinctive role. The amateurs were to proceed, like their naturalist counterparts, on a Baconian programme of fact-collecting, while the professionals would accept and utilise their data, together with the results obtained from their own more sophisticated instruments. This is a situation which, to some degree, obtains even today. In the early years of astrophysics Lankford argues that more important discoveries came from amateur observers than from established observatories in England, in contrast to both America and Germany.[34]

It is interesting to note that the first President of the BAA, Captain William Noble, was also a leader in the movement against State intervention in, and funding of, scientific research. A corollary of such a position was the high value placed on straightforward, simple and even individual research.

The importance of these events for the study of professionalisation is twofold. First, astronomers, like chemists, were caught up in the larger movements in the last thirty years of the nineteenth century, a period of critical importance for professionalisation as a whole in Britain. Everyone was aware of claims and counter claims by so-called 'professionals' in almost every walk of life. But on the other hand this is a cautionary tale for those disposed to generalise too widely. For what happened in 1890 was not the establishment of a professional institution, but of its polar opposite. Here it was not professionals but amateurs asserting their rights and autonomy. Nor, except perhaps in the case of State intervention, were the amateurs trying to put the clock back. Noble himself admitted that amateurs, like professionals, would need to specialise.[35] The difference between the chemists and the astronomers reminds us that, though both groups lived and worked within a common culture, having common values, there was more than one way of responding. Partly, of course, this was a matter of the differing perceptions as to the social utility of the two subjects. Chemistry was by now regarded as the most 'useful' of all sciences, while astronomy was probably the least useful. Yet as we have been reminded before utility is not always a matter of practical and economic consideration, and astronomy far more than chemistry had a great role in the arguments of natural theology which, even in the late nineteenth century, were still being heard in Britain. A more likely cause of the differences is to be found in the simple fact that the two sciences were, at least in Britain, at such different stages of internal development. It was quite appropriate for astronomers to engage in large scale fact-collecting. But in chemistry no theory had been as long established as (say) Newtonianism and even theories of valency and structure were undergoing considerable change.[36] In some cases basic premisses were being questioned. In Germany the production of thousands of new organic compounds might perhaps have approximated to the equivalent of the astronomers' fact-collecting. But it was otherwise in England, and so the scientific communities could, and did, react in different ways.

For subjects less rigorous than astronomy, coexistence of amateurs and full-time workers continued well into the twentieth century, while professionalisation in its full modern sense is incomplete to this day. Thus O'Connor and Meadows have argued that 'a separation of amateur and professional . . . occurred in geology only after the Second World War',[37] and then chiefly after quarrels about the kind of publications carried by the various national and regional institutions for geology.

Much the same situation arose with ecology.[38] Once this became generally recognised as 'scientific natural history', the gulf widened between the professional (who was 'scientific') and the amateur (who presumably was not). This differentiation also took place in the present century, through the gradual exclusion of the amateur from plant ecology which may perhaps be traced back to the division of labour proposed by the Hampstead naturalist Arthur Tansley in 1904. He proposed associations of workers who would characterise, enumerate and describe plants, but stressed the role of those who would require special training for the scientific analysis of results. Theirs was a 'much more difficult, as it is undoubtedly a much higher task, than the descriptive one'.[39] The emphasis on training and organisation networks represents a concept of professionalism much broader than mere full-time employment. Even that most descriptive of sciences, natural history, was contemplating such a prospect only four years after the close of the nineteenth century.

In conclusion, it is instructive to note what Robert Fox has called the 'creeping professionalisation' of French science in the second half of the nineteenth century.[40] This was the general extension of research-oriented academic positions for the 'softer' sciences, and with it the imposition on the practitioners of the 'professional' criteria of excellence, particularly in formal examinations and research publications. From the beginning of the century physics, mathematics and chemistry had been highly 'professionalised' in this sense. But natural history, geology and the life sciences, as in England, had flourished in the provincial societies, being pursued by laymen whose enthusiasm for fieldwork compensated for lack of specialised training and academic vocation.

By 1850 many French academics, especially those in science and medicine, earnestly wanted reform and a new emphasis on research. Conceivably this reflected their aspirations for special status, though it could also be argued that it caused it. Either way the links between professionalism and reforms in academia were extremely strong. Within the Université this 'professionalism' began to assert itself in the 1860s, two of its most visible spokesmen being the physiologist Bernard and the zoologist Lacaze-Duthiers. The intention was to raise the academic status of the life sciences by deliberately degrading fieldwork and arguing for laboratory-based research. As Fox remarks, this conception of academic respectability 'was resplendently enshrined in the research laboratories for zoology, palaeontology, and botany that were established in the science faculties from the 1880s'.[41] Certainly the dream became a reality under the Third Republic.

13. The State and Science

13.1 PRESSURES FOR GOVERNMENT SUPPORT

IT is curious that in the growing involvement of the state with science in Victorian Britain, a process that involved amongst other things the greater democratisation of science, a name that occurs at several crucial points is unquestionably aristocratic. The Dukes of Devonshire had immense wealth and much land, including the famous estate of Chatsworth. Their ancestry included two families that had produced distinguished men of science: the Hon. Robert Boyle and the Hon. Henry Cavendish. The sixth Duke, who died in 1858, is ironically remembered chiefly for his head gardener and estate manager, one Joseph Paxton (1803–65). This enterprising horticulturist, who had attained a national reputation by the 1830s, constructed two large conservatories at Chatsworth in which he even upstaged the Royal Gardens at Kew by feats of remarkable technical prowess.[1] On the basis of that experience he was able to design that pioneering structure of glass and iron which was to house the Great Exhibition in London of 1851. Rather than destroying many of the fine trees in Hyde Park, it was proposed to build a vast Exhibition Hall to enclose them.[2] The result was a building covering twenty acres with nearly one million square feet of glass supported by 3300 iron columns. The 'Exhibition of the works of industry of all nations' was conceived by the Prince Consort himself, but he was ably assisted by Henry Cole, secretary of the Royal

Society of Arts, of which the Prince was President. In 1850 Lyon Playfair was co-opted to help.[3] Exhibits were classed under (i) raw materials, (ii) machinery, (iii) manufactures and (iv) sculptures and the fine arts. Fountains played, organs pealed and above all competitions were held with juries of British and foreign members in equal numbers.[4]

By any standards, it was a phenomenal success. In 144 days it received over six million visitors, and made a clear profit of £186,436. Many medals were won by British industry, whose triumphs coincided with a growing national awareness of a connection between science and national prosperity. As recently as 1849 a reformed Royal Society had been granted a modest £1000 by the Government, and in 1850 that citadel of conservatism and classicism, the University of Oxford, created an Honours School in Natural Science (as well as one in Modern History). Inspired as much by the new geology of Lyell as by the German ideals of science as an indispensable part of culture, that notoriously elusive entity, British public opinion, was beginning to show a wider appreciation of science as a factor in national prosperity and a symbol of national prestige. The Great Exhibition confirmed the trend. Indeed it accelerated it, for now government was at last taking notice. The Queen's Speech for 1852 promised a comprehensive programme for art and science, and important developments may be traced directly back to the Exhibition, largely owing to the efforts of members of the Commission which had been set up to run it and which was continued as a permanent institution. In 1853 the Government set up a Department of Science and Art under the education office of the Privy Council. Cole and Playfair were joint secretaries. It took under its wing the School of Mines and the Museum of Practical Geology. With much of the Exhibition's profits, supplemented by a Government grant of £150,000, the Commissioners purchased eighty-six acres in South Kensington with a view to forming a national science centre. The vision was never completely fulfilled, although Imperial College, the Royal College of Art and the Museums are a partial embodiment of the optimistic hopes of the 1850s. A further 'achievement' of the Department of Science and Art was the notorious system of payment-by-results for science teaching, this being bureaucratically tidy but educationally disastrous. A comprehensive series of examinations in science and technology was instituted under its auspices in 1859. It included examinations for teachers, who would receive certificates attesting their competence.

Yet all was not well, even in 1851. In that year Charles Babbage published another blistering attack on the scientific establishment generally, concluding in words redolent of his 1830 polemics, 'science in England is not a profession; its cultivators are scarcely recognised even as a class'.[5]

The point has been made elsewhere that Babbage was guilty of over-statement, but he was entirely right to put his finger on the great inherent weakness of English science: the low standing and remuneration of its practitioners. That this was so became all too obvious a few years later, in the 1860s.

In 1861 the death of Prince Albert deprived English science of an important stimulus and of support in high places. Because of his death the next Exhibition was postponed till the following year; and it was on a much smaller scale than in 1851. It seemed to indicate that Britain was holding her own in terms of international competition, though in fact other nations were catching up. But within five years complacency was shattered and a new spirit of urgency, even despair, was induced by the Paris International Exhibition held in the Spring of 1867. Of the ninety departments, Britain received prizes in only ten. Understandably it was widely held that British industry had got itself into a lamentably low state. Lyon Playfair observed:[6]

> I am sorry to say that the opinion prevailed that our country had shown little inventiveness and made little progress in the arts of industry since 1862. . . . The one cause upon which there was most unanimity of conviction is that France, Prussia, Austria, Belgium and Switzerland possess good systems of industrial education for the masters and managers of factories and workshops, and that England possesses none.

But of all the European competitors it was Germany who was chiefly feared. Since the 1830s research had been seen as a normal feature of German universities. The high seriousness of their members was reinforced by local pride, for competition between the German states led to a proliferation of research laboratories and rivalry to attain the highest standards of work. This decentralisation of the university system was to continue to some extent even after the Imperial Government began to take over various technical projects towards the end of the century. Even Paris lost its supremacy as a place of scientific pilgrimage to a Germany which boasted universities like Göttingen, Jena, Berlin, Bonn, Freiberg, Breslau, Königsberg, Marburg, Giessen, Leipzig, Heidelberg, Halle, Münich and so on, at all of which some science was included, and where (in 1864) no less than 260 professorial staff were engaged in its teaching. Moreover the Technical Institutes in Germany were, in this period, attracting students at a faster rate than even the ordinary universities.[7] By 1870 Germany had about 2000 students reading science and technology, out of a total of around 18,000. Comparable figures to these were not obtained in Britain until thirty years later. Thus it was that Matthew Arnold was able to complain that French universities had no liberty; English universities had no science; but German universities had both.[8] An obsession with German technological and scientific supremacy in the 1860s was not confined to the English, and Robert Fox has stressed the general fears amongst French scientists and their admiration for the German system of university-based research.[9]

The British disquiet was expressed in two movements that eventually coalesced. The first was a powerful lobby for improved education in what we should call 'technology' but which the Victorians called 'art'.[10] Even if it was technology, rather than science, that chiefly profited, the movement for technical education was of immense importance for the future of all science in Great Britain. Of the dozens of immediate reactions to the 1867 Exhibition, and to Playfair's assessment of it, one only can be mentioned here: the case of John Scott Russell, Scottish shipbuilder and naval architect, colleague of Brunel, and a rather neglected figure in Victorian science. He had been one of the Commissioners for the 1851 Exhibition, and in 1869 published his *Systematic Technical Education for the English People.* A true agitator, he used his acquaintance with

Henry Cole to the utmost effect although, in the end, he could not persuade him to abandon his role of ideologically neutral bureaucrat. Russell proposed a 'Ministry of Public Education' headed by a man who had himself 'high education', and run by highly qualified scientists rather than 'official place men'. There should be a central, metropolitan technical university, with a supporting system of fifteen provincial technical colleges and a thousand trade schools. It would cost the exchequer one million pounds per annum. But, as Russell's biographer observes, 'he was just too early'.[11] A second, related movement, was that for 'the endowment of research'. Both of these concerns were taken up by an informal London-based group that had been in existence since 1864 and was poised to take any action it deemed necessary to improve the state of science in the country.

This was the so-called X-club.[12] It consisted of nine close friends who were all in leading positions in English science: George Busk (anatomist), Edward Frankland (chemist), T. A. Hirst (mathematician), J. D. Hooker (botanist), T. H. Huxley, (physiologist), John Lubbock (naturalist), Herbert Spencer (philosopher), William Spottiswoode (mathematician), and John Tyndall (physicist). All except Spencer were Fellows of the Royal Society. This informal association met 240 times, its members being united in their dedication to the ideals of scientific research, to the necessity for government support and to the need to pursue science independently of religious dogmas. They regularly discussed the affairs of more formal societies, suggested candidates for elections and made efforts to promote government and popular support for science. Not least important as an organ of their views was the journal *Nature*, founded in 1869 by Norman Lockyer who, though not a member of the Club, sympathised with many of its ideals. Although it appears they first met in the aftermath of the early Darwinian controversy, and in some fear of theological interference with science, they were well placed to exert the maximum pressure on behalf of movements that emerged at the end of the decade.

In response to the general public interest in the subject the Government established in 1870 a Royal Commission on Scientific Instruction and the Advancement of Science, under the chairmanship of the seventh Duke of Devonshire, and hence known as the Devonshire Commission.[13] In 1868 a paper had been presented to the British Association at Norwich by a retired Indian Army Officer, Alexander Strange. Recently returned to Britain, he was surprised at the lack of state support for scientific research, and the case presented in his paper for government help for science was so compellingly made that a British Association Committee was set up to look into the matter. It included Frankland, Hirst, Huxley, and Tyndall. The Committee urged a Royal Commission which, in due course came into being.[14] X-club members were also represented on this Commission (Lubbock and Huxley).

Between 1872 and 1875 the Commission produced eight major reports:

1. The Science School at South Kensington.
2. Technical Education.
3. Science at Oxford and Cambridge.
4. Museums and Public Lectures.
5. London and other University Colleges.
6. The Teaching of Science in Public and Endowed Schools.
7. Scottish Universities.
8. The Government and Science.

It was this final report that publicised the evidence put forward by Strange (and others), not least the proposal for a Ministry of Science which the Commissioners enthusiastically endorsed. Together with Lockyer and a few others, Strange took a radical view, that only an entirely new system of state support would provide both the institutions and the salaries needed by scientific research workers. From now on 'endowment of research' was a movement as well as a war cry.[15] Thus in 1876 a volume appeared of *Collected Essays on the Endowment of Research*, edited by Charles E. Appleton, Fellow of St John's College, Oxford, and an enthusiast for German science; most of the essays were written by Oxford men.

By the late 1870s considerable pressure had been built up for Government involvement in science, both in inspiring technical education and in providing state support for research. Before we consider the outcome it is as well to note that the cry for Government action was far from unanimous, and that other pressures were mounting to delay or eliminate the possibility of direct Government involvement.

13.2 OPPOSITION TO GOVERNMENT SUPPORT

Opposition to the endowment movement came from part of the scientific community as well as from those outside it. If we consider, first, the objections raised by those who might have been supposed to have had much to gain, practising scientists themselves, it is immediately apparent that a multiplicity of objections could be raised, and it is rash to assert that any one was 'more important' than the other. Furthermore, as Ruth Barton observes, the campaign in favour of state support simply cannot be interpreted in terms of the two main political parties,[16] and the same is true of the opponents.

First, there were those who simply regarded the campaign as unnecessary. Those historians who have argued that complicated scientific apparatus required Government finance would do well to consider the 'string and sealing wax' tradition that permeated chemistry, physics and the life sciences. Only in astronomy did the professional routinely need very expensive apparatus, and it is interesting that it is precisely among the astronomers that the most vocal opponents of Government support may be found. It was a simple fact that much valuable work could be done, and was being done, with simple equipment; and it was also a fact that private benefactions could produce extremely good results. After all it was in 1873 that the Cavendish Laboratory opened in Cambridge, with a benefaction from the seventh Duke of Devonshire of £8450. And there were still scientific researchers rich enough to finance their own laboratories, as was the case of Lord Rayleigh, who later succeeded Maxwell at the Cavendish. Of course laboratories were needed, and Frankland was extremely glad to move in to the new one in South Kensington at this time. But it is easy to exaggerate, and the spirit of self-help lived long into the late nineteenth century. Thus George Airy,

Astronomer Royal, observed in his presidential address to the British Association in 1851:

> the absence of Government-Science harmonises well with the peculiarities of our social institutions. In Science as well as in almost everything else, our national genius inclines us to prefer voluntary associations of private persons to organisations of any kind dependent on the State.[17]

He was saying much the same thing twenty years later.

It was also seen that State involvement might be difficult as well as unnecessary. Speaking of the Royal Society's Government grants, Airy claimed 'it is commonly acknowledged, even by supporters of research, that they were awarded in a "lax and unbusinesslike way" '.[18] Further, such a movement could be divisive. It is interesting that the bitterest controversy was amongst astronomers, one beneficiary at an early stage of the movement having been Norman Lockyer, who was virtually endowed with a solar physics laboratory at South Kensington. He was still involved in controversy of the strongest and most personal kind. Leading the attack was the astronomer Captain W. Noble, who became Honorary Secretary of a 'Society for Opposing the Endowment of Research' and who contributed (under the pseudonym FRAS) numerous letters and articles to the *English Mechanic*, which thus became a battleground for the endowment movement. He said:

> the 'endowment of research' has come in these later days to signify the subsidising of such things as 'committees on solar physics', and not in the very slightest degree the helping of the real student. It means the opening of the National Purse, the shutting of the Nation's eyes, and the seeing what Brompton will send us.[19]

The suggestion that Lockyer, who had a large and needy family, was feathering his own nest at the expense of the public purse, was hardly justified, but perhaps understandable in view of the personal rancour between Lockyer and some of his rivals. If, therefore, the possibility of State help was likely to produce divisions within the scientific community, it could hardly have come at a worse time. For it was exactly now that the movements for professionalisation and association were gaining momentum, when science was under threat from outside, and when a united front was much to be desired. It was all most unfortunate. Not surprisingly, the prospect of even large sums of money being distributed amongst the community was foreseen by many as having an overall deleterious effect.

Lastly, there were those who felt that Government endowment was positively dangerous. Airy certainly saw this:[20]

> I think that successful researches have in nearly every instance originated with private persons, or with persons whose positions were so nearly private that the investigators acted under private influence, *without incurring the dangers attending connection with the State.* Certainly I do not consider a Government is justified in endeavouring to force at public expense investigations of undefined character, and, at the best, of doubtful utility; and I think it probable that any such attempts will lead to consquences disreputable to science. The very utmost, in my opinion, to which the State should be expected to contribute, is exhibited in the large grant entrusted to the Royal Society.

Before them the scientific community had the unsavoury spectacle of the Ayrton affair,[21] in which Gladstone's autocratic First Commissioner of Works, A. S. Ayrton, became locked in a bitter dispute with Joseph Hooker, Director of the National Arboretum at Kew, over what was, fundamentally, 'who ran science'. Although the affair ended inconclusively it generated immense ill-feeling. The frustrating and embarrassing experience of Hooker was an example of how State interference could be extremely dangerous for the prosecution of disinterested research. This view of the endowment movement represents science, not as a means of social control, but as an object of it. And if one thing united members of the X-Club, and of many more formal organisations as well, it was a detestation of outside interference. That, no doubt, is why

X-Club members tended to take a pragmatic or middle view, seeking to gain some financial support from the State but jealously guarding their own freedom to pursue research in the way they wanted.

That, however, is only one side of the picture. Within the scientific community there was a belief that State involvement in science might be unnecessary, difficult, divisive or even dangerous. But outside there was opposition of a rather different kind. We may identify two strands of that opposition. First there was the development of a strong anti-science lobby in the years following the Darwinian controversy. Much of this was brought to a head by Tyndall's address[22] to the British Association in Belfast in 1874, in which he professed himself to be a 'materialist', speaking of 'the impregnable position of science' in contrast to 'all schemes and systems which thus impinge upon the domain of science', theological or otherwise. They must 'submit to its control'. While it is perfectly true that Tyndall's materialism was qualified, that he limited the subjugation of theological systems to those which actively opposed science, and that he did not launch a frontal assault on other ideological systems, the effect was predictably shocking. The British Association declined to print the address, and it was the subject of sermons in most Belfast churches the following Sunday. Whatever the merits or demerits of the position Tyndall was advocating, it must be confessed that for sheer political ineptitude his speech takes some beating. This is particularly so in view of his own strong association with others in the X-club who sought both to minimise Government control and to maximise its support. Few gestures could be more likely to have the opposite effect than his celebrated Address. But we must not blame Tyndall too much. Without doubt he represented a position held by a good many people though for the public at large it was a very unpopular attitude to take. The arrogance of positive science is always an unattractive thing, and it was particularly manifest in England at just the same time as the endowment of research movement was gaining strength. This was also suffering in some quarters from movements devoted to the abolition of vivisection, and others opposed to the practice of vaccination.

A second kind of obstacle to the endowment movement

came from the simple, if hard, facts of political life. Governments at this time were constantly occupied with other pressing matters, and such questions as the endowment of research were not seen to be high priorities. Palmerston's Government in the 1860s was especially concerned with matters of foreign policy, and in the 1870s few Governments were in office long enough to make much impact on the matter. There was also the paradox that the Liberals were headed by Gladstone who, for personal reasons, found little sympathy with science, and yet they probably had more scientists among their supporters than did the Conservatives. As Gladstone's biographer observed:

> From any full or serious examination of the details of the scientific movement he stood aside, safe and steadfast within the citadel of Tradition.[23]

On the other hand the Conservatives under Disraeli were less likely to be favourable towards science because of the political complexion of many of its supporters. In the event they did turn out to be the party more likely to be sympathetic, but they were frustrated by being put out of office in 1880 just when they might have contemplated legislation. And Roy MacLeod has drawn attention to the attitude of the Permanent Secretary to the Treasury, Ralph Lingen, who was strongly opposed to spending public money on scientific research, and wrote 'I cannot see what is the obligation of Government while the Royal Commissions are sitting on two universities with more than £700,000 per annum to dispose of'. He undoubtedly exerted an inhibiting influence on successive Governments in the 1870s and 1880s.[24]

13.3 THE CONCLUSION OF THE MATTER

Not surprisingly, Government response was as ambivalent as the attitudes of the scientific community that had elicited or opposed it. The endowment of Lockyer's solar physics work at South Kensington was part of a general tidying up and rationalising operation for the complex of institutions in that area. In 1876 the Royal Society original grant-in-aid of £1000 was supplemented by a new sum of £4000 which could be used for personal research funding. It was to be administered by a Government Fund Committee of the Royal Society, with money provided on a yearly tenure basis. In 1881 money spent over the previous quinquennium (£20,000) had given rise to 21 papers in *Philosophical Transactions* and 150 more in other journals.[25] In the last fifteen years of the century, while the grant remained at the same level, various fellowships were provided and research institutions founded or aided, such as the London School of Tropical Medicine, established in 1899 by the Colonial Office. It was not until the twentieth century that major improvements in the scale of Government support took place.

In view of frequent claims about the magnitude of the discrepancy in Government support for science between England and Germany, and possible consequences for the whole political and economic history of the twentieth century, it may be as well to conclude with one or two cautionary remarks. It is not true that science received no patronage from the Government until the last three or four decades of the nineteenth century. W. H. Brock has identified no less than six areas of State patronage at an earlier stage, which suggests that the picture painted by Babbage was much too pessimistic. There was, in the first place, the chemical research undertaken at the Royal Mint when, between 1850 and 1870, assaying of coinage metals was undertaken by men who for part of their time were able to undertake chemical research. Secondly, the Inland Revenue and Excise Department had employed scientists from time to time to look into specific problems such as denaturing of alcohol, and other problems associated with excise duty. Assistants were sent to study at appropriate institutions in order to get a wide scientific background. Then the Admiralty and War Department spent money on maintaining the health of service personnel (thus subsidising the pharmacists who supplied drugs) and promoted the study of explosives at Woolwich Arsenal. In medicine such matters as sanitation, water purification and sewage disposal were regularly undertaken by consultant chemists such as Frankland, Hofmann and Playfair, who could therefore be said to have received Government subsidy for their labours. In the area of education the Department for Science and Art between 1853 and 1870 spent nearly £52,000 on its examinations. Finally, some support for scientists has been identified in pensions money paid to them. None of this individually amounts to very much, but collectively it portrayed considerable Government interest in science, even though that is quite unquantifiable.[26] One may also identify the more hidden Government investment in science as part of its programme of colonial expansion, demonstrated particularly in its patronage of the Royal Botanic Gardens at Kew and other overseas analogues which it absorbed from the East India Company. Voyages of discovery, the transferring of cinchona trees from South America to India and of rubber trees from Brazil to plantations in India, Ceylon and elsewhere in the East, were as much fruits of Government initiated scientific research as of its policies of imperial expansion.[27]

Even when Government support did exceed these limits, and could be seen as a gesture by the State in favour of disinterested and independent scientific research, it is still not true that it was an unmixed blessing. As Roy MacLeod has written:

> The endowment of research movement helped to create the impression that scientists could be identified by certain unifying attitudes and could be considered a collective 'community' within society. . . . In a sense, the 'researchers' victory was a paradox. The public acceptance that brought them recognition in universities and research institutes also helped separate them from industry and society at large.[28]

So the professionalism that had demanded support paid the price of increasing isolation.

It is instructive to notice that parallel situations existed outside England. Robert Fox has decisively demonstrated that, on the negative side, the doldrums into which French science fell in the middle of the nineteenth century are not (as often supposed) due to deficient government attitudes, but the result of such considerations as the centralisation of French higher education, its preoccupation with the classics and other more personal factors.[29]

When France and Germany came into conflict in 1870 the role of governments in science became a matter of urgent public debate, especially in the defeated France.[30] The great Louis Pasteur, so appalled by his country's humiliation that he returned to the University of Bonn the Diploma for an Honorary Doctorate awarded three years earlier,[31] attributed the French defeat to a neglect 'of the great labours of thought for half a century, particularly in the exact sciences'.[32] Doubtless he was reflecting on his own recent battle with the authorities for better laboratory facilities. During the war science had been applied to food supply, medicine, explosives manufacture and communications technology.[33] Yet the German victory owed far more to good organisation and the uptake of foreign (especially British) technology than to German scientific research. Nevertheless, for good reasons or bad, the question of science and government control continued to be popularly debated in France. Two weaknesses, in particular, were pinpointed. The chemist J. B. Dumas complained of the dangers of centralisation:

> How can a town with a University, receiving from Paris its administrators, professors, budget, programme and diplomas for its students, be actively interested in its prosperity? Is it not evident that its degree of co-operation and initiative will be determined by its responsibility? Its municipal authority, and the country's leaders, regard the establishment of French higher education as a thing for the state; in other countries it is for the town. We could cite the instance of the University of Basle, which is at our very gates, where masters, pupils, inhabitants, are united in a common interest like a single family, and passionately follow the progress of the ancient and illustrious institutions which honour their city.[34]

His colleague Deville blamed bureaucracy:

> It is said on all sides, and justly, that we have been defeated by science. The cause lies in the regime that has burdened us

> for eighty years, a regime that deals with scientific affairs, its dissemination, teaching and application by bodies or offices which lack competence, and consequently the love of progress.[35]

He could have said much more. During the recent conflict the famous Académie, which contrived to continue its regular Monday meetings even in a city under siege, gave consideration to scientific aspects of the military arts, – always provided suggestions had come from its own members. Contributions by outsiders had first to be vetted by committees. And even if there is an element of charade and unreality about the Académie at this time, it is even more notable that in conversations with the government it was the Académie which took the initiative.[36]

From 1879 to about 1902 a temporary renaissance in French science has been identified, and attributed to educational reforms inspired by the German precedent.[37] In the Third Republic a programme of radical educational reforms placed much emphasis on science, partly as an instrument of national recovery and partly as an ideological weapon in a positivist crusade against reaction, whether civil or ecclesiastical. Spearheading the advance was the French chemist M. Berthelot.

A major aim of this programme was decentralisation, as Dumas had urged. And it was partially successful, most notably at Toulouse where work of real relevance to the regional economy was accompanied by academic research of the future Nobel Laureate P. Sabatier on catalytic hydrogenation of organic substances.[38]

Towards the end of the nineteenth century French academic scientists had addressed themselves to national economic needs, especially where financial autonomy was granted by the administration. In 1890 an Institut Chimique was opened in Nancy with 500,000 francs from the government and an equal amount from local councils. Private finance was also encouraged and business men in French provinces made numerous grants to support research which might have economic benefits for themselves and their community.[39]

But what of Germany in all this period? To judge from

contemporary British allusions, Germany was the paradigm of a state which gave proper support to science. Not least was this true in education, and of the multitude of English voices eulogising the German system and arguing for its emulation in Britain, that of Matthew Arnold must speak for them all.[40]

It is true that German higher education had a long established infrastructure of technical schools, many of which had become colleges and even the prestigious Technical High Schools, all of which were from the first supported by the state. Prior to 1871 individual German states had vied with one another for excellence in technical education. And that, to a greater extent than in France or England, involved practical training for industry. An expatriate German, having lived in England for many years, was the alkali chemist George Lunge. Addressing the Newcastle Chemical Society as President he urged:

> The principle upon which the teachers of chemistry in the German Universities and Polytechnic Colleges . . . proceed is that the budding chemists shall receive a perfectly sound theoretical knowledge. It is considered that a time of two years is rather too little – that three years ought to be the minimum for the scientific training of a chemist; but after that they are most decidedly encouraged to go through a course of practical instruction; and that principle having now been acted upon for some years past, the result has been that there are at this moment a very large number of chemists in Germany who combine scientific training with practical knowledge; . . . that combination alone . . . will produce proper results.[41]

The deep and characteristic commitment to research which marked most German universities had fostered the tradition of research Ph.Ds decades before the conflict with France. Yet caution is necessary in interpreting this as an unrelieved blessing. The very size of university departments could prove a stumbling block, as Wilfred Farrar remarked:

> Education through research can, unless great care is taken . . . result in deep but narrow knowledge, and the blinkered

specialist, the semi-literate Ph.D., was already a problem figure in Germany by the 1880s.[42]

State patronage of science was not limited to education. For half a century Prussian officials had been encouraged to travel, particularly to Britain, and engage in what can only be termed, however impolitely, industrial espionage.[43] Such traditions died hard after the war, though in the nature of the case statistics are not easy to find.

The constitution of 1871 laid upon the Imperial Government only a limited commitment to science, chiefly in respect of weights and measures, patents, coinage and professional licensing (especially for medicine and pharmacy). However, in practice the government went a good deal further. In a discussion of scientific organisation and science policy in Imperial Germany from 1871,[44] F. Plesch has identified three phases of its activity. For the first few years it was concerned simply to continue administering the institutions taken over from the North German Confederacy, especially Prussia. These included various institutions for military research.

A second phase, the late 1870s and 1880s, concerned the development of science for economic ends. During this period the proportion of expenditure on economic to military research rose from 0.2:1 to 1.28:1. Funding went to institutions concerned with sea, air, biological resources, as well as with standardisation and testing. In the third phase (from the late 1880s) it is argued that the intensification of economic, political and cultural activities in Germany led to a great awareness of competition with the outside world, and a consequent establishment of scientific activities in Africa and other German colonies.

This analysis would lend support to the popular image of state-run science. However, certain important qualifications must be made. First, as Plesch admits,[45] Imperial Government policy tended to favour 'directed' science (in contrast to that of some individual German states).

A second consideration is more serious, and takes into account the crucial role of certain individuals in the development of research institutions. A classic case is the

Physikalich-Technische Reichsanstalt, the first national institution for pure and applied scientific research.[46] Opened in Charlottenberg, Berlin, in 1887, it arose from a need for precision technology, as perceived by numerous scientists in the 1870s, including DuBois-Reymond and Helmholtz. Even more influential was the industrialist Werner Siemens, who doggedly pursued the project despite much opposition for fifteen years, laid out half a million marks of his own money in order to buy the land, and canvassed unceasingly for further support throughout the country. It is hard to see how the story of the P.T.R could justify a legend of complete state support. In a recent paper[47] David Cahan has drawn attention to the vacillations of the Imperial Government, and to Siemens' assessment of Bismarck as someone who 'holds science for a type of sport without practical meaning'.[48] He concludes that industrial and political needs were of secondary importance in the founding of the P.T.R. 'Far more important were Werner Siemens' personal desires for the advancement of pure science, the securing of scientific foundations for technology and his patriotism'.[49] Once again one sees the importance of the gifted individual which must be set against alleged influences of impersonal economic forces or of science policy. Cahan adds 'neither Prussian nor Reich "science policy", if, as is most doubtful, any such policy existed, played any role in the establishment of the P.T.R.'.[50]

A third way in which the picture of monolithic state support needs to be modified is to consider the support given by industry. It has recently been argued that government support was by no means the only factor in assuring the growth of science, and that in Germany an underestimated cause has been the strongly motivated, and well endowed pharmaceutical industry, supporting research back in the eighteenth century and continuing into the nineteenth.[51] Research reports of individual chemical firms later in the nineteenth century confirm that this trend had continued.[52]

Certainly Germany differed from her European neighbours in many respects. Her scientific education, unlike that in England, was state supported from the first; and her scholars, unlike many in France, enjoyed a high degree of academic freedom. And there were other differences also. Writing of the

French synthetic dyestuffs industry two authors have concluded 'at the end of the nineteenth century, university research and industrial research in France formed two separate worlds which scarcely touched each other. It was very different in Germany'.[53] On the other hand it is possible to concentrate too greatly on matters of institutional arrangement. Peter Lundgreen has argued that if a 'study is confined to the field of physics and if indicators such as productivity per scientist or total investment per scientist are taken into account, the differences between France and Germany are negligible around 1900'.[54] And as a cautionary tail to his paper he argues:

> Apart from the question of 'heroes' (great and small ones) in economic history, a *mésalliance* of Mercury and Minerva in France, in contrast with their supposedly idyllic marriage in Germany, is hard to verify.[55]

14. Epilogue: Science Triumphant?

On Monday, 12 August 1844, the pioneer of the chemical atomic theory, John Dalton, was honoured in death by his fellow citizens. Against the wishes of his Quaker friends (and probably his own as well) he was given a public funeral in Manchester, after a lying-in-state in a gas-lit room in the Town Hall, where it is said over 40,000 people came to pay their last respects. Through streets lined with silent crowds the procession of nearly one hundred carriages made its way to Ardwick Cemetery, where his remains were laid to rest. Most remarkably, in an almost unprecedented gesture, shops and warehouses throughout the city were closed in honour of the great natural philosopher. For many years Dalton had rarely left the city of his adoption, and the pageantry of his funeral testifies to the vigour of local patriotism. Here was a man of learning, whose atomic theory was little heeded by his scientific contemporaries and almost certainly incomprehensible to most of his fellow Mancunians, and yet he had in some way helped to put their city on the map. They were very proud of him.[1]

Nearly forty years later another hero of science was eulogised in a display of public grief. On 26 April 1882 the body of Charles Darwin was conveyed from the Chapel of St Faith, where it had lain over night, to its final resting place within the nave of Westminster Abbey, that national shrine for kings, queens, conquerors, national leaders and heroes of every kind. It lay within a few feet of the memorial to Sir Isaac Newton himself. This time the pomp and pageantry was on a national, rather than a local, scale. Just as local dignity had brushed aside the wishes of the Quaker community in which Dalton had lived, so now national pride over-ruled the desires of Darwin's neighbours for a burial in his local village of Downe in Kent.[2] Each of the funerals represents in an extreme form society's determination to honour a man of science, but it is surely not being too cynical to assert that it also exemplifies society's determination to extract the maximum benefit even from his parting. Yet in other respects the two events were very different and marked a profound change in the social appreciation of science over those turbulent and critical years between 1844 and 1882.

In Dalton's time the chief social use for science had been to minister to local civic pride and to be seen as a mark of provincial culture. The death of Dalton was thus a considerable blow to Manchester as a whole. But by the 1880s the cultural functions of science had changed to something considerably more complex. And it is not too much to say that this change had been engineered in a quite deliberate and skilful manner.

To read the eulogies of science that fill the popular literature of late Victorian England is to see science as the beneficial agent that had transformed the quality of life in almost all its departments and which had been doing so since at least the Industrial Revolution. In fact, however, it is exceedingly difficult to justify this view before about 1850. Science had for several decades before then been employed by a few of the most far-sighted industrial entrepreneurs, but mainly in connection with their own immediate objectives, and its perceived effects on ordinary life were small. If sources before that time are consulted, one finds that science certainly has a high 'kudos', but not on account of its immediate practical effects. Instead it is seen in the context of natural theology as supporting and undergirding the claims of revealed religion. In other words it is seen as an ally of the Church. It was one ingredient of what Robert Young has called the 'common intellectual context',[3] and it harmonised (more or less) with all the other ingredients. Difficulties about the age of the earth and the extent of Noah's flood were seen as relatively minor problems, rather than as major confrontations. By the 1870s, and certainly the 1880s, all of this had gone and science had assumed a new social role, one that included a powerful antagonism to religion. The broadest description of this new role may be said to be secularisation, its agent being Victorian Scientific Naturalism – the view that nature's activity can be interpreted without recourse to God, spirits etc., i.e. without bringing in the *super*natural. In skilled hands it was a most potent weapon and it may be considered as the direct cause of a major social change that began in the late nineteenth century, the secularisation of society.

Why then did the social role of science undergo this metamorphosis? It is tempting to ascribe all to Darwin. His *Origin of Species* of 1859 and subsequent writings removed the keystone of natural theology (the argument from design) and replaced it with a notion of evolution by 'natural' selection, in other words denying any role in the biological world to supernatural selection or any other kind of divine activity. That is how the picture was often painted in the late nineteenth century and how it is often presented today. But this is far too simple, and failed to account for many awkward facts. Darwin himself – certainly in 1859 – was reluctant to deny the possibility of God's activity 'behind the scenes'; a significant number of committed Christians (like Charles Kingsley and Henry Drummond) saw no necessity to abandon their faith and in fact enthusiastically embraced Darwinism; very many scientists of the first rank either rejected evolution for a variety of reasons or simply ignored it, (such as Kelvin and Maxwell). In any case there was no logical reason why evolution and natural selection could not be fitted within a theistic framework, but the theory could be assailed on other grounds and did (and does) offer an easy target for those who are spoiling for a fight and wish, for other reasons altogether, to attack it.

Darwinism itself is not sufficient to explain the growth of scientific naturalism and therefore secularisation. After about 1850 science was beginning to make a much more visible impact on the life of ordinary people and thus could be appealed to with much more confidence by those who sought to exalt its claims. From Chadwick's Sanitary Report of 1842, through the activities of the Medical Officers of Health to the Public Health Act of 1875, the public was becoming increasingly aware of the potential role of science in improving public health. Even so in 1875 the mortality rate was virtually the same as it had been a quarter of a century before, though in the next twenty-five years it fell dramatically. At first chemical and then biological analysis of water supplies, together with new methods of purification and an understanding of the aetiology of many infectious diseases, led to insights into what was necessary for a healthy population to exist under urban conditions.

In the country, repeal of the Corn Laws in 1846 necessitated much more efficient agricultural practice, and the land-owners became aware of this and of the possibilities of more scientific farming. Again the immediate returns were small, but by the end of the century it was abundantly clear that the application of science could at least make possible much greater yield of crops and therefore more abundant food. Again the foreign competition, especially as revealed by the Exhibitions of which the Victorians were so proud, led to a swift realisation that if Britain was to maintain its place in world markets it simply had to take more seriously the application of science to manufacture. Developments in metallurgy led to new engineering wonders (such as the Forth Bridge) while the applications of chemical science produced such diverse materials as celluloid, aluminium and a dazzling variety of synthetic dyestuffs. Coincident with these advances, the scientific community in Britain was becoming far more specialised and (in places) professionalised. The regrets of a Babbage or a Brewster that science did not bring adequate rewards were finding more powerful and widespread expression within the institutions of science. A triumphalist view of science was much to be encouraged for financial and institutional reasons, quite apart from any alleged threat from theology. That such a threat was perceived in some quarters simply gave a peculiar twist to the argument. Science was presented as being all-triumphant, not merely over an obdurate nature or an all-prevailing ignorance, but quite specifically over certain aspects of Christian theology.

Yet one wonders whether it really was quite as simple as that. That many people of this time lost a formal commitment to Christian belief cannot be doubted for one moment; but how often was this because of a genuine confrontation with natural science? Young has cited cases of the secularist movement, sermons of country parsons and debates among the ecclesiastical Lords and concluded that scientific ideas, if mentioned, were not critical to any of the arguments.[4] More often than not, as with the experience of Edward Frankland who lost his youthful religious beliefs, it was a reading of Tom Paine that caused the renunciation.

If the writings of members of the X-Club are examined, one thing emerges with great clarity. All the members adopted an anti-religious stance and all were deeply resentful of any question of religious interference in the practice of science. Here, then, we have a clue. The problem was not how old was the earth, or Genesis versus Geology, but wherein was to lie the cultural leadership of the new generation of intellectuals. Up to this point the Church had exerted great direct or indirect control over education and cultural leadership throughout national life. The aim of the secularising scientists was therefore to displace the ecclesiastical hegemony with a scientific one. If this interpretation is accepted, many other features of Victorian life fall into place, and it becomes apparent that science was being used as an ideological weapon to advance the interests of the community that it founded.[5] It was only natural, therefore, to flood the market with accounts of the achievements of science and of the 'heroes' who had brought them about. It was natural to produce such works as J. W. Draper's *History of the Conflict between Religion and Science* (1875) or A. D. White's *The History of the Warfare of Science with Theology in Christendom* (1895) in which organised religion (Draper) or dogmatic theology (White) were each shown as the victims of an inexorable tide of advancing science. Such Whiggish historiography would not be acceptable today, but it served well the cause of triumphalist science. Indeed one is not too far from the mark to assert that the conflict thesis, (or the view that science and religion are best understood historically in terms of the conflict between them) is an artifact of Victorian social ambition.

But if science were to become a substitute for religion, and to displace Christianity as the dominant mode of thinking in British thought, it was desirable to retain certain of its cultural features rather than jettison them all, for in this way it would be more apparent that science had really taken over the leadership. Thus we find, particularly in the writings of Huxley, references to the 'church scientific' and even to himself as 'bishop'. Ecclesiastical styles in architecture were taken over by the secularising scientists; one has only to look at the great halls of many Victorian universities, complete with Gothic windows, soaring arches and great organ. Better still one may survey the Natural History Museum of South Kensington, 'nature's cathedral'. Most poignantly of all Charles Darwin is virtually canonised by his burial in Westminster Abbey. In so far as secularisation of society may be seen to be a significant social change it becomes important to examine the role of science in the arguments of many (but by no means all) scientists of late Victorian England.

Finally, the sacralisation of science (or the secularisation of society by putting science in the place of cultural leadership once occupied by institutional religion) has meant a burgeoning of the uses of science as an argument for justifying, or delaying, changes in society. Most famously, the theory of evolution has been widely used (or misused) in political propaganda. But as David Oldroyd has shown,[6] all 'shades of political opinion were able to draw support from the Darwin/Wallace theory', in agreement with G. B. Shaw's remark that Darwin 'had the luck to please everybody who had an axe to grind'. To analyse such 'uses' of science in the late nineteenth century and beyond, or its more direct effects on almost every part of human society, would need another book. One may well agree with the opinion expressed by Richard Westfall that 'much of the modern world appears . . . as so many epiphenomena to the growth of science'.[7]

Notes and References

Place of publication is London unless otherwise stated.

1. APPROACHING THE PROBLEM

1. It is perfectly possible in earlier periods to find anticipations of modern views which seem to call their modernity in question. Many of these anticipations were still-born, or else can be shown on more careful examination to have meant something very different from their apparent equivalents at a later stage. Two obvious examples are Greek views of a heliocentric solar system and eighteenth-century speculations on 'evolution'. They do not alter the transformations of the overall picture that took place, and no attack of 'precursoritis' (a disease once prevalent amongst some historians of science) should blind us to the magnitude of such changes.

2. R. R. Palmer, 'The World Revolution of the West, 1763–1801', *Political Science Quarterly*, 64 (1954) 1–14.

3. R. M. Young, 'Science *is* social relations', *Radical Science Journal*, no. 5 (1977) 65–129.

4. D. M. Knight, *The Nature of Science* (Andre Deutsch, 1976) p. 11.

5. M. Hammerton, 'A fashionable fallacy', *New Scientist*, 76 (1977) 274–5.

6. That is, the law that heat never of itself passes from a cold body to a hotter body.

7. D. W. F. Hardie, *Business History*, 8 (1966) 734 (review of R. E. Schofield's *The Lunar Society of Birmingham*).

8. John Durant, 'The Hammerton Thesis – a reply', *New Scientist*, 76 (1977) 485.

9. The papers were reprinted in *Science at the Cross Roads*, with a foreword by J. Needham and a new introduction by P. G. Werskey, (Frank Cass, 1971).

10. M. Berman, ' "Hegemony" and the amateur tradition in British Science', *J. Soc. Hist.*, *1* (1975) 30–50 (30).

11. G. N. Clark, *Science and Social Welfare in the Age of Newton*, (Oxford: Clarendon Press, 9th edn., 1949), pp. 86–9.

12. A. Thackray, 'Natural knowledge in cultural context: the Manchester Model', *Amer. Hist. Rev.*, 79 (1974) 672–709. He points to 'an almost unnerving scholarly unanimity' in locating within industry the roots of Mancunian science, but associates it with an 'enduring ignorance of sources'.

13. Berman, ' "Hegemony" and the amateur tradition in British Science'.

14. It is well-known that Marx and some of his followers appear to deny an absolute distinction between society and nature. See, for example, Robert Young, 'The historiographic and ideological contexts of the nineteenth-century debate on man's place in nature', in M. Teich and R. Young (eds.), *Changing Perspectives in the History of Science* (Heinemann, 1973) 344–438 (especially 428–35). One's response to that theoretical approach will depend on how far it is possible to understand and agree with the presuppositions behind it, and how far it appears to be conformable to experience in the widest sense of the term. It is perfectly possible to reject totally the philosophical/political stance involved and still to be grateful for insights which have been developed from it which may, however, find their legitimation from quite different sources.

15. See, for example, G. N. Cantor and M. J. S. Hodge, *Conceptions of Ether: Studies in the history of ether 1740–1900* (Cambridge: Cambridge University Press, 1981).

16. F. A. Kekulé, *Berichte*, 23 (1890) 1306.

17. See, for example, C. A. Russell, *Coal, the basis of Nineteenth Century Technology*, Unit 4 of the Open University Course AST 281, 'Science and the Rise of Technology since 1800' (Milton Keynes: Open University Press, 1973; corrected reprint 1976).

18. Thackray, 'Natural knowledge in cultural context', 676.

19. A. Schuster and A. E. Shipley, *Britain's Heritage of Science* (Constable, 1917) p. xiii.

20. B. Barnes and S. Shapin (eds), *Natural Order: Historical Studies of Scientific Culture* (Beverley Hills and London: Sage Publications, 1979).

21. M. Rudwick, *Hist. Sci.*, 18 (1980) 269–85 (270).

2. THE SHAPE OF ENLIGHTENMENT SCIENCE

1. T. Sprat, *The History of the Royal Society of London for the Improving of Natural Knowledge* (1667), p. 113.

2. A vast literature exists on the Scientific Revolution. General treatments of wide scope include: Marie Boas, *The Scientific Renaissance, 1450–1630* (Collins, 1962); A. R. Hall, *From Galileo to Newton, 1630–1720* (Collins, 1963); R. S. Westfall, *The Construction of Modern Science: Mechanisms and Mechanics* (Cambridge: Cambridge University Press, 1977); A. R. Hall, *The Scientific Revolution, 1500–1800* (Boston, Mass.: Beacon Press, 1957).

3. J. Thomson, 'The Seasons': 'Summer', ll, 1560–3, (1746).

4. A. Pope, 'Epitaph intended for Sir Isaac Newton' (1730).

5. A. Pope, 'An Essay on Man', Epistle II, ll. 31–4 (1734).

6. See, for example, R. E. Schofield, 'An evolutionary taxonomy of eighteenth-century Newtonianisms', *Studies in eighteenth century culture*, 7 (1978) 175–202; M. C. Jacob, 'Newtonianism and the origins of the Enlightenment: a Reassessment', *Eighteenth Century Studies*, 2 (1977) 1–25; I. B. Cohen, *The Newtonian Revolution with illustrations of the transformations of scientific ideas* (Cambridge: Cambridge University Press, 1980). On Newton himself there is an immense literature, including the following full-length biographies: D. Brewster, *The Life of Sir Isaac Newton* (1836); L. T. More, *Isaac Newton* (New York: Scribners, 1934); E. N. da C. Andrade, *Sir Isaac Newton* (Collins, 1954); F. E. Manuel, *A Portrait of Isaac Newton* (Belknap Press of Harvard University Press, 1968); R. S. Westfall, *Never at Rest. A Biography of Isaac Newton* (Cambridge: Cambridge University Press, 1980).

7. J. Swift, *Gulliver's Travels*, Part 3, Ch. V.

8. Cited in C. E. Raven, *Natural Religion and Christian Theology* (Cambridge: Cambridge University Press, 1953) p. 124.

9. Hall, *Galileo to Newton*, p. 320.

10. For a general account of scientific developments in this period see: A. Wolf, *A History of Science, Technology and Philosophy in the Eighteenth Century*, 2nd edn rev. by D. McKie, (Allen & Unwin, 1952); A. Ferguson (ed.), *Natural Philosophy through the Eighteenth Century* (Taylor and Francis, 1948) (reprints of articles about the *Philosophical Magazine* from 1948); G. S. Rousseau and Roy Porter (eds), *The Ferment of Knowledge: Studies in the Historiography of Eighteenth-Century Science* (Cambridge: Cambridge University Press, 1980).

11. See I. B. Cohen, *Franklin and Newton: an Inquiry into Speculative Newtonian Experimental Science and Franklin's Work in Electricity as an Example thereof* (Philadelphia: American Philosophical Society, 1956); J. L. Heilbron, *Electricity in the Seventeenth and Eighteenth Centuries: a Study of early Modern Physics* (Berkeley, Los Angeles and London: University of California Press, 1979).

12. A traditional view expressed with great force in A. D. White, *A History of the Welfare of Science with Theology in Christendom* (Arco Publishers Ltd., 1955) vol. 1, p. 366. The work, first published in 1895, is a classic expression of the now outmoded 'Conflict Thesis' which sees science and religion (or theology), as essentially opposed, with triumphant science inevitably the victor.

13. B. Franklin, *Autobiography, to which is added Jared Sparks' Continuation (abridged)* (Hutchinson, 1903) pp. 241 and 248. See also C. van Doren, *Benjamin Franklin* (New York: Viking Press, 1938).

14. Sir Henry Lyons, *The Royal Society, 1660–1940. A History of its Administration under its Charters* (Cambridge: Cambridge University Press, 1944) p. 193.

15. Ibid, pp. 193–4.

16. Hall, *The Scientific Revolution*, p. 352.

17. van Doren, *Franklin*, p. 171.

18. T. S. Kuhn, *The Structure of Scientific Revolutions* (Chicago, London: University of Chicago Press, 1962) pp. 21–2.

19. See, for example, M. P. Crosland, 'Chemistry and the Chemical Revolution', in Rousseau and Porter, *The Ferment of Knowledge*, pp. 389–416.

20. H. Butterfield, *The Origins of Modern Science 1300–1800* (Bell, 1950) Ch. XI.

21. J. J. Berzelius, *Traité de Chimie* vol. IV (Paris, 1831) p. 554.

22. See, for example, W. M. Sudduth, 'Eighteenth-century identifications of phlogiston with electricity', *Ambix*, 25 (1978) 131–147.

23. On Lavoisier see, for example, D. McKie, *Antoine Lavoisier, Scientist, Economist, Social Reformer* (Constable, 1952); H. Guerlac, *Antoine-Laurent Lavoisier, Chemist and Revolutionary* (New York: Charles Scribner's Sons, 1975); and W. A. Smeaton, 'New light on Lavoisier: the research of the last ten years', *Hist. Sci.*, 2 (1963) 51–69.

24. A. Thackray, in D. S. L. Cardwell (ed.), *John Dalton and the Progress of Science* (Manchester: Manchester University Press, 1968) Ch. VI.

25. See C. A. Russell, G. K. Roberts and N. G. Coley, *Chemists by Profession* (Milton Keynes: Open University Press, 1977) Ch. II.

26. Cited in C. Singer, *The Earliest Chemical Industry: an Essay in the Historical Relations of Economics & Technology illustrated from the Alum Trade* (Folio Society, 1948) pp. 196 and 214, from Jars' *Voyages Métallurgiques* (Paris, 1774–81) vol. III.

27. See E. C. Cripps, *Plough Court, the Story of a notable Pharmacy* (Allen and Hanburys Ltd., 1927).

28. A. Raistrick, *Quakers in Science and Industry* (Newton Abbot: David and Charles, 1968) p. 284.

29. Cited in A. C. Wootton, *Chronicles of Pharmacy* (Macmillan, 1910) vol. II, p. 188.

30. A. Kent 'The Place of Chemistry in Scottish Universities', *Proc. Chem. Soc.*, (1959) 109–113 (110).

31. Raven, *Natural Religion*, p. 157.

32. Ibid., p. 147.

33. See A. O. Lovejoy, *The Great Chain of Being, A Study of the History of an Idea* (Cambridge, Mass.: Harvard University Press, 1953).

3. NATURE IN THE ENLIGHTENMENT

1. D. Hartley, *Observations on Man*, (1749).
2. J. D. Bernal, *Science in History*, 3rd edn (Penguin, 1969) vol. II, p. 509.
3. Techniques may of course reflect progress and interest, as in the first appearance of a geological hammer (1696) or a butterfly net (1711), D. E. Allen, *The Naturalist in Britain* (Penguin Books, 1978) p. 6, but they do not necessarily do so.
4. T. H. Levere, 'Friendship and Influence: Martinus van Marum, F.R.S.', *Notes and Records 25* (1970) 117.
5. James Thomson, 'The Seasons', 1st edn (1726) lines 301–17.
6. John Davy, *Memoirs of the Life of Sir Humphry Davy, Bart.* (1836) vol. I, p. 51.
7. Thomas Gray, 'Elegy in a Country Churchyard' (1742) 1, 1.
8. D. Diderot, *Pensées sur l'interprétation de la Nature*, (Paris, 1754), Section XVII.
9. W. Derham, *Physico-theology*, 4th edn (1716) p. 368, 80.
10. T. Burnet, *The Sacred Theory of the Earth*, 3rd edn (1697) vol. I, p. 95.
11. P. Townby and L. Whibley (eds), *Correspondence of Thomas Gray*, vol. III, 1766–1771 (Oxford: Clarendon Press, 1935) p. 1088, (Letter to Wharton, November 1769).
12. Burnet, *The Sacred Theory of the Earth*, p. 97 similarly in 6th edn (1726) p. 194.
13. D. Defoe, *A Tour through the whole Island of Great Britain* (1724–6) (Everyman Edition, 1962) vol. II, p. 269.
14. M. H. Nicolson, *Mountain Gloom and Mountain Glory: the Development of the aethetics of the infinite* (Ithaca, N.Y.: 1959) p. 345.
15. Derham, *Physico-Theology*, p. 71.
16. J. Addison, *The Spectator*, no. 412 (23 June 1712).
17. Paul Shepard, *Man in the Landscape: A Historic View of the Esthetics of Nature* (New York: Knopf, 1967) especially pp. 119–56.
18. E. W. Hodge believed that 'underlying social and economic reasons permitted, though they could hardly shape', the change in attitudes that led to tourism. But he did concede that 'the establishment of such a tradition of sightseeing requires the interaction of two factors at least: the actual practice of local guides, and the more academic endorsement of people laying claim to "taste".' (E. W. Hodge, 'The Monstrous Impending Craggs', *Journal of the Fell and Rock Climbing Club of the English Lake District,* 10 (1935) 179–94 (183, 185).)
19. See A. R. B. Haldane, *New Ways through the Glens* (Newton Abbott: David and Charles, 1962).
20. Folio map by Bowles and Sayers from *The Royal English Atlas* (1760).
21. E. W. Hodge, *Enjoying the Lakes* (Edinburgh: Oliver and Boyd, 1957) p. 19.
22. B. Willey, *The Eighteenth-century Background* (Penguin, 1972) p. 162.
23. John Evelyn, a Trustee, believed that 'theist' meant 'deist' – E. de Beer, *The Diary of John Evelyn* (Oxford, 1955) vol. V, p. 88.
24. R. E. W. Madison, *The Life of the Honourable Robert Boyle, F.R.S.* (Taylor and Francis, 1969) p. 274.
25. Richard Bentley, 1662–1742, then a young clergyman and chaplain to the Bishop of Worcester, became eventually Master of Trinity College, Cambridge and one of England's great classical scholars. His Boyle Lecture appeared in print as *A Confutation of Atheism from the Origin and Frame of the World* (1693).
26. Letter 10th December 1692, published in *Four Letters from Sir Isaac Newton to Dr Bentley* (1756) p. 1.

27. M. Hunter in *Science and Society in Restoration England* (Cambridge: Cambridge University Press, 1981) p. 186, argues that Ray's traditional design argument was as influential as Newtonianism.

28. Robert Boyle, 'A free enquiry into the vulgarly received notion of nature' in M. B. Hall (ed.), *Robert Boyle on Natural Philosophy* (Indiana: Indiana University Press, 1965) 150.

29. H. Butterfield, *The Origins of Modern Science 1300–1800* (Bell, 1949) p. 150.

30. Humphry Davy, notebook 13E (undated but before 1800), at the Royal Institution, London.

31. Robert Boyle, *Usefulness of Experimental Natural Philosophy* (1663).

32. Samuel Clarke, 'A discourse concerning the unchangeable obligations of natural religion and the truth and certainty of the Christian revelation' (1705) in R. Watson (ed.) *A Collection of TheologicalTracts* (1785) vol. IV, p. 248.

33. S. Clarke, *An Exposition of the Church Catechism* (1799) pp. 249–250 (my italics).

34. *The Works of the Reverend John Wesley* vol. XI (1812) p. 403.

35. J. D. Bernal, *Science and History*, vol. II (Penguin, 1969) p. 515; in view of this author's previously expressed view of the low state of science at that time, such an attitude might in any case have been expected!

36. Karl Barth, *From Rousseau to Ritschl*, trans. H. H. Hartwell (SCM Press, 1959) p. 18.

37. C. H. Faust, 'Jonathan Edwards as a scientist', *American Literature*, 1 (1930) 393–404.

38. A. A. Dallimore, *George Whitefield: the Life and Times of the great Evangelist of the Eighteenth-century Revival* (Banner of Truth Trust, (vol. I) 1970 and (vol. II) 1980 especially vol. II, pp. 441–53).

39. *The Works of the Rev. John Wesley* (1809) vol. IV, 389, 344 and 427; vol. III 162.

4: NATURE AND SOCIETY

1. S. Clarke, (ed.), *Sermons on Several Subjects* (1715) pp. 122, 120.

2. William Lloyd, *A Sermon Preached Before the Queen*, (30 January 1691) p. 29.

3. *The Works of . . . Joseph Butler* (Oxford: 1844) vol. II, p. 281.

4. D. C. Douglas, (Gen ed.), *English Historical Documents*, A. Browning (ed.), vol. VIII, 1616–1714 (Eyre and Spottiswoode, 1953) p. 126.

5. Newton's thoughts on providence have been extensively studied: See for example, F. Oakley, 'Christian theology and the Newtonian science: the rise of the concept of the laws of nature', *Church History*, 30 (1961) 449; J. E. McGuire, 'Force, active principles, and Newton's invisible realm' *Ambix*, 15 (1968) 154–208; D. Kubrin, 'Newton and the cyclical cosmos; providence and the mechanical philosophy', *J. Hist. Ideas*, 28 (1967) 325–346.

6. A. Dyce, (ed.), *The Works of Richard Bentley* (1838) vol. III, p. 22.

7. The *Spectator*, 13 June 1712, no. 404.

8. See, for example, R. Hooykaas, *Religion and the Rise of Modern Science*, corr. edn (Edinburgh and London: Scottish Academic Press, 1973).

9. M. C. Jacob, *The Newtonians and the English Revolution, 1689–1720* (Hassocks: Harvester Press, 1976) p. 177; see also H. Guerlac and M. C. Jacob, 'Bentley, Newton and Providence (The Boyle Lectures once more)', *J. Hist. Ideas*, 30 (1969) 307–318.

10. This point was made more strongly and fully in a paper by Dr E. Duffy given recently at a conference of the British Society for the History of Science. He argues not only that 'Latitudinarianism' is a term void of useful meaning, the examples quoted by Jacob varying widely in their views of important issues. He also points out that

'Newtonianism' in the hands of its theological exponents was actually subversive, rather than supportive, of the post-Revolutionary church settlement (several Boyle Lecturers being known as dangerous heretics, notably Whiston and Clark). I am grateful for being allowed to see a copy of Dr Duffy's manuscript in advance of its publication.

11. Jacob, *The Newtonians*, p. 60.

12. A. Napier, (ed.), *The Theological Works of Issac Barrow* (Cambridge: 1859) vol. v, pp. 231–2.

13. Cited in Jacob, *The Newtonians*, p. 14.

14. R. Watson, *A collection of Theological Tracts* (1785) vol. IV, p. 143.

15. Ibid., p. 175.

16. C. Wordsworth, (ed.), *The correspondence of Richard Bentley* (1842) vol. I, p. 39.

17. Jacob, *The Newtonians*, pp. 178–179.

18. Ibid., 179.

19. Ibid., p. 173.

20. Ibid., p. 174.

21. Ibid., p. 160.

22. Ibid., p. 51.

23. Napier, *Works of Isaac Barrow*, 9, p. 578.

24. Jacob, *The Newtonians*, p. 174. In just such a manner John Ray justified the wisdom of God in making noxious insects; one might 'as well upbraid the prudence and policy of a state for keeping forces which generally are made up of very rude and insolent people, which yet are necessary, either to suppress rebellions, or punish rebels or other disorderly and vicious persons, and keep the world in quiet' (John Ray, *The Wisdom of God Manifested in the Works of Creation*, 9th edn (1727) p. 375).

25. Jacob, *The Newtonians*, p. 180. The Derham passage is from his *Physico-theology: or a demonstration of the Being Attributes of God, from His Works of Creation* (1714) pp. 32–3.

26. See G. Holmes, *B. J. Hist. Sci.*, 11 (1978) 164–71 (168), for a review of Jacob, *The Newtonians*.

27. R. E. Schofield, *Mechanism and Materialism: British Natural Philosophy in an Age of Reason* (Princeton: 1970) p. 21.

28. Jacob, *The Newtonians*, p. 174.

29. Ibid., p. 168.

30. Ibid., p. 173.

31. G. Holmes, *B. J. Hist. Sci.*, p. 169.

32. C. C. Gillispie, *Genesis and Geology* (New York: Harper, 1959) p. 20.

33. Ibid.

34. R. Porter, *Social History*, 3 (1978) 249.

35. F. E. Manuel, *The Religion of Isaac Newton* (Oxford: Clarendon Press, 1974) p. 38.

36. Ibid., p. 34.

37. H. Guerlac, 'Further Newtonian Studies', *Hist. Sci.*, 17 (1979) 75–101 (82–5).

38. For an excellent account of this see Jacob, *The Newtonians*, ch. IV; see also her 'Newtonianism and the Origins of the Enlightenment: a Reassessment', *Eighteenth Century Studies*, 11 (1977) 1–25.

39. See, for example, A. D. Orange, 'Oxygen and one God: Joseph Priestley in 1774', *History Today*, 24 (1974) 773–781.

40. Two recent accounts of this are G. N. Cantor, 'Revelation and the Cyclical Cosmos of John Hutchinson', in L. J. Jordanova and Roy Porter (eds), *Images of the Earth* (Chalfont St Giles: B.S. Hist. Sci., 1979) pp. 3–22; and C. B. Wilde, 'Hutchinsonian Natural Philosophy and Religious Controversy in Eighteenth Century Britain', *Hist. Sci.*, 18 (1980) 1–24.

41. Alan Richardson, *The Bible in the Age of Science* (SCM., 1961) p. 38.

42. D. Greene, 'Augustinianism and empiricism. A note on Eighteenth Century English Intellectual History', *Eighteenth Century Studies*, 1 (1967) 33–68.

43. Wilde, 'Hutchinson Natural Philosophy', p. 20.

44. A. Thackray, *Atoms and Powers* (Cambridge, Mass.: Havard University Press 1970) p. 247.

45. Ibid., pp. 280 ff.

46. Michaud's *Biographie Universelle* (Paris: 1818) vol. XXI, p. 84.

47. Charles Wesley's hymn 'And can it be . . . ', (1738) (my italics).

48. R. E. Schofield, 'John Wesley and Science in Eighteenth Century England', *Isis*, 44 (1953) 331–340.

49. *The Works of the Rev. John Wesley* (1808) vol. IV, p. 261.

50. Tract printed in *Works of John Wesley.*, vol. XI, p. 403. (Cf. p. 47).

51. *Meditations among the tombs; reflections on a flower garden; a descant on creation* (first published 1746).

52. *Contemplations on the night: Contemplations on the starry heavens: a winter piece* (first published 1747).

53. J. Ryland, *The Character of Rev. James Hervey, M.A.* (c. 1791) p. 144. On Hervey see also J. C. Ryle, *Christian Leaders of the Eighteenth Century* (1885) pp. 328–57 (Banner of Truth Trust reprint 1978). Hervey, like Priestley, earned the disapproval of Dr Johnson, whose 'meditation on a pudding' (1773) was a parody of Hervey's extravagant prose (taking also a side swipe at the views of the 'theoretical Burnet'); G. B. Hill (ed.), L. F. Powell, (rev. edn), J. Boswell, *Life of Johnson*, 2nd edn (Oxford: Clarendon Press, 1964) vol. V, pp. 351–2.

54. Ryland, 'Hervey' p. 306; so Hervey says in a letter to Whitefield.

55. James Hervey, *Meditations and Contemplations* combined volume (1820) pp. 430–431.

56. Ibid., p. 240.

57. Ibid., pp. 303–4.

58. Ibid., p. 311.

59. Ibid., pp. 100–201.

60. Ibid., p. 410.

61. Ibid., p. 244.

62. Ibid., p. 161.

63. Ibid., p. 301.

64. Ibid., pp. 460-461.

65. Ibid., p. 157.

66. Ibid., p. 332.

67. Ibid., p. 331, 324.

68. Ibid., p. 370n.

5. THE SOCIAL ORGANISATION OF ENLIGHTENMENT SCIENCE

1. Modern full length treatments range from Sir Henry Lyons, *The Royal Society 1660–1940* (Cambridge: Cambridge University Press, 1944) to Marjory Purver, *Royal Society: Concept and Creation* (Routledge and Kegan Paul, 1967); many other books, for example, B. J. Shapiro, *John Wilkins 1614–1672* (Berkeley, Los Angeles and London: University of California Press, 1968) refer extensively to the circumstances of its creation, and there are innumerable papers, especially in *Notes and Records of The Royal Society*.

2. The definitive work is Roger Hahn, *The Anatomy of a Scientific Institution: The Paris Academy of Sciences, 1666–1803* (Berkeley, Los Angeles and London: University of California Press, 1971).

3. Hahn, *Anatomy of a Scientific Institution*, pp. 43–4.
4. Humphry Davy, Royal Institution Notebook 13H (undated but early 1800s).
5. *Histoire de l'Académie Royale des Sciences et Belle Lettres (de Berlin)* 24 (1767) 362–4; cited in Hahn, *Anatomy of a Scientific Institution*, p. 44.
6. Hahn, *Anatomy of a Scientific Institution*, p. 50.
7. The Royal Society *Journal Book* cited in Lyons, *The Royal Society*, p. 24.
8. Lyons, *The Royal Society*, pp. 166–68.
9. This was the case with Benjamin Franklin, according to a letter to his son dated 19th December 1767. He discovered in old Council books of the Society a reference to his own admission, where he learned that 'the honour being voluntarily conferred by the Society, unsolicited by me, . . . my name was entered on the list with a vote of Council, *that I was not to pay anything*. (*The Autobiography of Benjamin Franklin*, new edn (Hutchinson, 1903) p. 367.)
10. Lyons, *The Royal Society*, p. 342.
11. Lyons, *The Royal Society*, p. 186.
12. T. H. Levere, 'Friendship and influence: Martinus van Marum', *Notes and Records*, 35 (1969) 113–120.
13. H. T. Parker, 'French administrators and French scientists during the old regime and the early years of the revolution', in H. T. Parker and R. Herr (eds) *Ideas in History* (Durham, N. C., 1965) 87–109 (90).
14. Hahn, *Anatomy of a Scientific Institution*, pp. 72–73. The Académie was an upholder of scientific orthodoxy, tending to exclude rival views such as Paracelsianism which, however, flourished in France outside the Académie (A. G. Debus, 'The Paracelsians in Eighteenth Century France', *Ambix*, 28 (1981) 36–51 (50)).
15. Lyons, *The Royal Society*, p. 215; pp. 144–5.
16. Hahn, *Anatomy of a Scientific Institution*, p. 63.
17. W. E. Knowles Middleton, *The Experimenters: A Study of the Accademia del Cimento* (Baltimore and London: Johns Hopkins University Press, 1971).
18. In her *Leisure and Pleasure in the Eighteenth Century* (Cassell, 1970) it is a pity that Stella Margetson almost entirely ignores this facet of English life.
19. Angus Armitage, *Edmund Halley* (Nelson, 1963) pp. 17 et seq.
20. D. E. Allen, *The Naturalist in Britain* (Penguin, 1978), p. 36.
21. Michaud, *Biographie Universelle* (Paris: 1825) vol. 43, p. 431n.
22. C. B. Boyer, *A History of Mathematics* (New York, London, and Sydney: John Wiley, 1968) pp. 501–2.
23. F. W. Gibbs, 'Bryan Higgins and his circle', *Chem.*, 1 (1965) 60–5; reprinted in A. E. Musson (ed.), *Science, Technology and Economic Growth in the Eighteenth Century* (Methuen, 1972) pp. 195–207.
24. F. W. Gibbs, *Joseph Priestley* (Nelson, 1965) p. 51.
25. Allen, *Naturalist in Britain*, p. 35.
26. For some of these books see D. M. Knight, *Zoological Illustration* (Folkstone: Dawson, 1977).
27. A. Burton, *Josiah Wedgwood* (Deutsch, 1976) pp. 49, 107, 112.
28. On Cavendish see D. C. Goodman in R. C. Olby (ed.), *Late Eighteenth Century European Scientists* (Oxford: Pergamon Press, 1966) pp. 95–126.
29. Boswell, *Life of Johnson*, (see note 53 to ch. 4) vol. I, pp. 140, 436; vol. II, p. 155 (Sulphuric acid reference, 1772); vol. III, p. 398.
30. 'Notices and anecdotes of Literati, collectors, &c. from a MS. by the late Mendes da Costa, and collected between 1747 and 1788', *Gentleman's Magazine*, 82 (1812) (1), 205–207 and 513–517.
31. One may cite the activities of the botanist/explorer Richard Spruce, whose collection of feathers from exotic South American parrots and other birds was made into an exquisite covering for a sofa which he presented to Lady Carlisle at Castle

Howard, on whose estate he had been born. (*Guide book to Castle Howard*, second edn. (Castle Howard Estate Ltd., 1974) p. 18.)

32. Allen, *Naturalist in Britain*, pp. 30–31, citing Joan Evans, *Pattern*, (1931).

33. P. M. Horsley, *Eighteenth Century Newcastle* (Newcastle: Oriel Press, 1971) pp. 88–9.

34. da Costa, 'Notices', pp. 513 and 205.

35. Ibid., p. 207.

36. Ibid., p. 206.

37. Allen, *Naturalist in Britain*, pp. 10–11.

38. Ibid., p. 19.

39. Ibid., p. 17.

40. *Critical review*, 1763, cited in Allen, ibid., p. 45.

41. See W. V. Farrar 'Science and the German University System, 1790–1850', in M. P. Crosland (ed.), *The Emergence of Science in Western Europe* (Macmillan, 1975) pp. 179–192; there is an excellent account of 'Germany before 1815', on pp. 179–182.

42. Ibid., p. 181.

43. Here in the Medical Faculty C. E. Weigel invented the mis-called 'Liebig condenser' in 1771: J. R. Partington, *A History of Chemistry* (Macmillan, 1964, vol. IV, p. 301.

44. Karl Hufbauer, 'Social Support for Chemistry in Germany during the Eighteenth Century: how and why did it change?', *Hist. Stud. Phys. Sci.*, 3 (1971) pp. 205–231 (218). See also his *The Formation of the German Chemical Community (1720–1795)* (Berkeley, Los Angeles and London: University of California Press, 1982).

45. W. Oberhümmer, 'Science in One City: the Vienna Story', *M & B Lab. Bull.*, 9 (1971) 88–91 (89).

46. Schemnitz was a town in West Hungary, now known as Banska Stiavnica, Czechoslovakia. This was almost the only institution in Europe to teach practical chemistry to large numbers of students: its first Professor of Chemistry, Mining and Metallurgy was N. J. E. von Jacquin, another Leyden man who eventually became Professor of Chemistry and Botany in Vienna. The laboratory was cited by Fourcroy as the model for the École Polytechnique of Paris. In 1786 it was the venue for an international meeting on a new process for the amalgamation of gold. See Partington, *History of Chemistry* (1962) vol. III, p. 146; F. Szabadvary, 'The Early History of Chemistry in Hungary', *J. Chem. Educ.*, 40 (1963) 46–48; notes by W. Oberhümmer and J. Zemplen, in D. S. L. Cardwell (ed.), *John Dalton and the Progress of Science* (Manchester: Manchester University Press, 1968) pp. 287–9; D. M. Farrar, 'The Royal Hungarian Mining Academy, Schemnitz: some aspects of technical education in the eighteenth century', (Manchester M.Sc. thesis, 1971).

47. In Saxony. The professor of mineralogy was A. G. Werner, founder of 'Neptunist' geology, to whom students flocked from all over the world.

48. See A. E. Nordenskjöld, 'A leaf from the history of Swedish natural science', *Nature*, 21 (1880) 518–21, 539–41, 563–5; also S. Lindroth, *Swedish Men of Science 1650–1950*, trans. B. Anderson (Stockholm: Almqvist and Wiksell, 1952), and *A History of Uppsala University 1477–1977,* trans. N. Tomkinson and J. Gray (Stockholm: Almqvistand Wiksell, 1976).

49. On all these see Partington, History of Chemistry, vol. III; also L. B. Hunt, 'Swedish Contributions to the Discovery of Platinum', *Platinum Metals Review*, 24 (1980) 31–39.

50. See R. Taton (ed.), *Enseignement et Diffusion des Sciences en France au XVIIIe Siècle* (Paris: Hermann, 1964).

51. See H. R. Trevor-Roper, 'The Scottish Enlightenment', *Studies on Voltaire and the Eighteenth Century* 58 (1967) 1635–58.

52. For an excellent account, to which this section is much indebted, see J. R. R.

Christie, 'The Origins and Development of the Scottish Scientific Community, 1688–1760', *Hist. Sci.*, 12 (1974) 122–141.

53. Ibid., p. 135.

54. J. B. Morrell, 'The Edinburgh Town Council and its University, 1717–1766', in R. G. W. Anderson and A. D. C. Simpson (eds), *The Early Years of the Edinburgh Medical School* (Edinburgh: Royal Scottish Museum, 1976) pp. 46–65.

55. A. Clow, 'Hermann Boerhaave and Scottish Chemistry', in A. Kent (ed.) *An Eighteenth Century Lectureship in Chemistry* (Glasgow: Jackson, 1950) pp. 41–48, (43).

56. Christie, 'Origins and Development of the Scottish Scientific Community', p. 128.

57. Clow, 'Hermann Boerhaave', p. 42. See also E. A. Underwood, *Boerhaave's Men at Leyden and After* (Edinburgh University Press, 1977); R. W. Innes Smith, *English Speaking Students of Medicine at Leyden* (Edinburgh and London: Oliver and Boyd, 1932).

58. Clow, 'Hermann Boerhaave', p. 44.

59. See R. L. Emerson, 'The Philosophical Society of Edinburgh', *B.J.Hist.Sci.*, 12 (1979) 154–191; 14 (1981) 133–176.

60. Christie, 'Origins and Development of the Scottish Scientific Community', p. 134.

61. S. Shapin, 'Property, Patronage and the Politics of Science: the founding of the Royal Society of Edinburgh', *B.J.Hist. Sci.*, 7 (1974) 1–41, (11).

62. J. Kendall, 'The First Chemical Society, The First Chemical Journal, and the Chemical Revolution', *Proc.Roy.Soc.Edinburgh*, 63A (1952) 346–359: 385–400

63. D. McKie, 'Some Notes on a Students' Scientific Society in Eighteenth-Century Edinburgh', *Science Progress*, 49 (1961) 228–241.

64. Kendall, 'The First Chemical Society', p. 394.

65. McKie, 'A Students' Scientific Society', p. 231.

6. SCIENCE IN THE EARLY INDUSTRIAL REVOLUTION

1. Amongst the vast literature dealing with this subject there may be mentioned T. S. Ashton, *The Industrial Revolution 1760–1830* (Oxford: Oxford University Press, 1947, reprinted 1970); E. J. Hobsbawm, *Industry and Empire*, vol. III of the Pelican Economic History of Britain (Penguin Books, 1969); D. S. Landes, *The Unbound Prometheus* (Cambridge: Cambridge University Press, 1970).

2. See, however, A. Thackray, 'Science and Technology in the Industrial Revolution', *Hist.Sci.*, 9 (1970) 76–89 (with special emphasis on Manchester).

3. A. R. Hall, *From Galileo to Newton, 1630–1720* (Collins, 1963) p. 332.

4. The most comprehensive treatment of this theme is still A. and N. L. Clow, *The Chemical Revolution* (Batchworth Press, 1952).

5. It was chiefly on the basis of chemistry that A. E. Musson and E. Robinson argued the case for a close link between science and industry in their paper 'Science and Industry in the late Eighteenth Century', *Econ. Hist. Rev.*, 13 (1960) (2). 222–244.

6. See H. W. Dickinson, 'The History of Vitriol Making in England', *Trans. Newcomen Soc.*, 18 (1937) 43–60; A. and N. L. Clow, 'Vitriol in the Industrial Revolution', *Econ. Hist. Rev.*, 15 (1945) 44–55.

7. Despite its name it contains no copper and is in fact hydrated iron (II) sulphate $FeSO_4.7H_2O$; the reaction may be approximately represented as:

$$2FeSO_4 + H_2O \rightarrow Fe_2O_3 + H_2SO_4 + SO_2.$$

8. It is in fact an acid (and a very strong one), an oxidising agent, a dehydrating agent, a sulphonating agent and a source of sulphate ions. The multiplicity of function accounts not only for the versatility of the material in effecting desired chemical

changes, but also the long period of time before its various functions could be unravelled and its chemistry properly understood. This did not happen until well into the nineteenth century.

9. At first it was a general reagent, used to make other chemicals; in diluted form it found limited application in pharmacy. Some of its later uses are referred to in the text: bleaching, iron pickling, chlorine and alkali manufacture, production of fertilisers, dyes, explosives and drugs. Gabriel Jars, *Voyages Métallurgiques* (Paris, 1781) vol. III, p. 309, visited English factories in the 1750s and 1760s and records that only Welsh-speaking women were employed, presumably to maintain secrecy! But there would be little point in this since Ward patented his process in 1749.

10. A. and N. L. Clow, 'John Roebuck', *Chem. and Ind.*, 61 (1942) 497–8.

11. Ibid., p. 497.

12. F. Home, *Experiments on Bleaching* (Edinburgh: 1756).

13. A. Clow, 'Chemistry in Scotland, and its Pioneer Contributions to Textile Technology', *J. Textile Inst.*, 52 (1961) 204–218.

14. J. Liebig, *Familiar Letters in Chemistry*, trans. J. Gardner (1843) p. 31.

15. A. and N. L. Clow, 'Lord Dundonald', *Econ. Hist. Rev.*, 12 (1942) 47–58.

16. A. and N. L. Clow, 'George Mackintosh (1739–1807) and Charles Mackintosh (1766–1842)', *Chem. and Ind.*, 62 (1943) 104–6.

17. Sal ammoniac, or ammonium chloride, was used for making ammonia and in other ways. See A. and N. L. Clow, 'Dr James Hutton and the manufacture of sal ammoniac', *Nature*, 159 (1947) 425.

18. R. E. Schofield, *The Lunar Society of Birmingham* (Oxford: Clarendon Press, 1963); E. Robinson, 'The Lunar Society and its Membership', *Trans. Newcomen Soc.*, 35 (1962–3) 153–177.

19. A. R. Hall, 'The Royal Society of Arts: Two Centuries of Progress in Science and Technology', *J. Roy. Soc. Arts*, 122 (1974) 641–658.

20. Whitehurst, 1779, quoted in Schofield, *The Lunar Society*, p. 177.

21. Watt, c. 1796, reproduced in E. Robinson and A. E. Musson, *James Watt and the Steam Revolution* (Adams and Dart, 1969) p. 204.

22. Schofield, *The Lunar Society*, passim.

23. Ibid., pp. 293–4.

24. Ibid., p. 237.

25. Ibid., p. 239.

26. Ibid., pp. 250–306.

27. Ibid., p. 228.

28. On the Manchester Lit. and Phil. see note 12 to Ch. 10 and also note 34 (below).

29. T. S. Ashton, *The Industrial Revolution 1760–1830* (London, Oxford, New York: Oxford University Press, 1970) pp. 16–17.

30. A. E. Musson and E. Robinson, 'Science and Industry in the late Eighteenth Century', *Econ. Hist. Rev.*, 13 (1960) 222–244. This has been expanded in the same authors' *Science and Technology in the Industrial Revolution* (Manchester: Manchester University Press, 1969).

31. Ibid., p. 244.

32. Ibid., p. 224.

33. Letter dated 1 February 1789, cited in Musson and Robinson, 'Science and Industry', p. 226 (italics mine). John Dalton did not join the Society until 1793.

34. A. Thackray, 'Natural Knowledge in a Cultural Context: The Manchester Model', *Amer. Hist. Rev.*, 79 (1974) 672–709; He claims that 'an adequate understanding of the society hinges on the question of the social legitimation of marginal men, on the adoption of science as the mode of cultural self-expression by a new social class, and on generational patterning in intellectual life' (p. 678); an

alternative interpretation of the Society's role will be considered in chapter 10. Thackray's analysis, particularly in its concept of 'marginality' has been attacked by I. Inkster (British Society for the History of Science *Newsletter* no. 8 (1982) pp. 15–18).

35. R. S. Watson, *The History of the Literary and Philosophical Society of Newcastle-upon-Tyne 1793–1896* (1897).

36. Watson, *History of the Newcastle Lit. and Phil.*, pp. 30–31, 41.

37. See G. A. Hedley, *Reports of the Literary and Philosophical Society of Newcastle* (Newcastle: 1823).

38. Watson, pp. 136–139; 339–341.

39. E. Robinson, 'The English "philosophes" and the French Revolution', *History Today*, 6 (1956) 116.

40. Schofield, *The Lunar Society*, pp. 134–139.

41. Ibid., p. 358.

42. W. V. and K. R. Farrar, and E. L. Scott, 'Thomas Henry (1734–1816)', *Ambix*, 20 (1973) 183–208.

43. Thackray, 'Natural Knowledge in a Cultural Context', p. 690.

44. Watson, *History of the Newcastle Lit. and Phil.*, pp. 33–41.

45. L. T. C. Rolt, *James Watt* (Batsford, 1962) p. 109.

46. W. V. and K. R. Farrar and E. L. Scott, 'Thomas Henry', p. 189.

47. W. V. and K. R. Farrar and E. L. Scott, 'The Henrys of Manchester Part II. Thomas Henry's sons; Thomas, Peter and William', *Ambix*, 21 (1974) 179–207.

48. P. M. Horsley, *Eighteenth-Century Newcastle* (Newcastle: Oriel Press, 1971) pp. 197–217; see also T. R. Knox, 'Thomas Spence: The Trumpet of Jubilee', *Past and Present*, 76 (1977) 75–98.

49. H. Lonsdale, *The Worthies of Cumberland* (1873) pp. 186–196.

50. Watson *History of the Newcastle Lit. and Phil.*, p. 46.

51. S. D. Chapman, *The Early Factory Masters* (Newton Abbott: David and Charles, 1967) p. 197. Chapman suggests that at these Societies political issues were not banned.

52. A. J. Turner, *Science and Music in Eighteenth-Century Bath* (Bath: University of Bath, 1977). List of members of the Bath Philosophical Society, p. 83.

53. H. Torrens, 'Geological Communication in the Bath Area in the last half of the Eighteenth Century', in L. J. Jordanova and R. S. Porter, (eds), *Images of the Earth: Essays in the History of the Environmental Sciences* (Chalfont St Giles: British Society for the History of Science, 1979) pp. 215–247.

54. S. Schaffer, 'Herschel in Bedlam: Natural History and Stellar Astronomy', *B.J.Hist. Sci.*, 13 (1980) 211–239 (213–4).

55. H. Hartley, *Humphry Davy* (Nelson, 1966) pp. 18–37.

7. INTERLUDE: CHANGE AND CONTINUITY IN FRENCH SCIENCE

1. For the first three sub-titles in this chapter I am indebted to Roger Hahn, 'The Problems of the French Scientific Community, 1793–1795', in *Actes du XIIe Congrès International d'Histoire des Sciences* (Paris, 1968, 1971), vol. IIIb, pp. 37–40.

2. H. T. Parker (note 4) dates this back to at least the historical writings of Alexis de Tocqueville of the nineteenth century.

3. See, for example, M. P. Crosland (ed.), *The Revolutionary Era described by Thomas Bugge* (MIT Press, 1969); C. C. Gillispie, *Science and Polity in France at the end of the old Regime* (Princeton University Press, 1980); L. P. Williams, 'Politics of Science in the French Revolution', in M. Clagett (ed.), *Critical Problems in the History of Science* (Madison: University of Wisconsin Press, 1962) 291–308.

4. H. T. Parker, 'French Administrators and French Scientists During the Old

Regime and the Early Years of the Revolution', in R. Herr and H. T. Parker (eds), *Ideas in History* (Duke University Press, Durham, N.C., 1965) pp. 85–109.

5. R. Hahn, *The Anatomy of a Scientific Institution: The Paris Academy of Sciences, 1666–1803* (University of California Press, 1971). Much of the information on the Académie depends upon this source.

6. C. Webster, Decimalisation under Cromwell', *Nature* 229 (1971) 463.

7. W. A. Smeaton, 'Decimalisation: The Origins', *Student Technologist*, 5 (1972) 22–3.

8. W. A. Smeaton, *Fourcroy, Chemist and Revolutionary, 1755–1809* (Cambridge: privately published, 1962) pp. 120–1.

9. At one stage the Secretary Condorcet had to communicate in duplicate with the Assembly and the King! (Hahn, *Anatomy of a Scientific Institution*, p. 166.)

10. Cited in Hahn, ibid., p. 227.

11. Parker, 'French Administration and French Scientists', p. 108.

12. S. Court, 'The *Annales de Chimie*, 1789–1815', *Ambix* 19 (1972) 113–128.

13. D. I. Duveen, 'A. L. Lavoisier and the French Revolution', *J. Chem. Educ.*, 31 (1954) 60–65.

14. Condorcet, Élote for Franklin, 30 November 1790, cited in Hahn, *Anatomy of a Sicentific institution*, p. 165.

15. L.-A.-G. Bose, 14 July 1790, cited in Hahn, *Anatomy of a Scientific Institution*, p. 263.

16. Hahn, ibid., p. 264.

17. R. R. Palmer, 'The World Revolution of the West, 1763–1801', *Political Science Quarterly*, 64 (1954) 1–14 (10).

18. T. Carlyle, *The French Revolution* (1888 edn) vol. II, p. 309.

19. D. W. F. Hardie, 'N. Leblanc', *Chem. Age*, 77 (1957) 208; see also J. G. Smith, *The Origins and Early Development of the Heavy Chemical Industry in France* (Oxford: Oxford University Press, 1979).

20. R. Glover, *Peninsular Preparation* (Cambridge: Cambridge University Press, 1963) pp. 63–7.

21. R. Watson, *Anecdotes of the life of Richard Watson, Bishop of Llandaff*, (1817) p. 149; L. J. M. Coleby, 'Richard Watson: Professor of Chemistry in the University of Cambridge, 1764–71', *Ann. Sci.*, 9 (1953) 101–23.

22. W. Miles, 'Early American Chemical Societies', *Chymia*, 3 (1950) 95–113 (99–100).

23. W. A. Smeaton, 'Chemical Education in Revolutionary France', *J. Roy. Inst. Chem.*, 82 (1958) 650–6; L. P. Williams, 'Science, Education and the French Revolution', *Isis*, 44 (1953) 311–30.

24. Hahn, *Anatomy of a Scientific Institution*, p. 278.

25. Smeaton, 'Chemical Education', p. 653.

26. On this see especially J. Langins, 'The Decline of Chemistry at the École Polytechnic (1794–1805)', *Ambix*, 28 (1981) 1–19, and references therein.

27. Smeaton, 'Chemical Education', p. 652.

28. Hahn, *Anatomy of a Scientific Institution*, p. 290.

29. Fourcroy, 1794, cited in Hahn, *Anatomy of a Scientific Institution*, p. 291.

30. Hahn, ibid., 290.

31. M. P. Crosland, 'The Congress on Definitive Metric Standards, *1789–1799*, The First International Scientific Conference?' *Isis*, 60 (1969) 226–231 (230).

32. M P. Crosland, 'The Development of a Professional Career in Science in France', *Minerva*, 13 (1975) 39–57 (40).

33. Hahn, *Anatomy of a Scientific Institution*, pp. 300–01.

34. Crosland, 'Development of professional Career in Science in France', pp. 44–5. How far French science was truly 'professionalised' depends on one's definition

(see Ch. 12). According to J. Ben-David the professionalisation of science is usually most effective in the second or third generations; the 'failure' of French science after 1830 therefore casts doubt on its professionalisation. He sees the new institutions as mainly a continuance of eighteenth-century patterns, temporarily interrupted by the Revolution (J. Ben-David, 'The rise and decline of France as a scientific centre', *Minerva*, 8 (1970) 160–79).

35. L. P. Williams, 'Science, Education and Napoleon I', *Isis*, 47 (1956) 369–382.

36. C. Babbage, *Reflections on the Decline of Science in England* (London, 1830) pp. 25–6.

37. D. Outram, 'Politics and Vocation: French Science, 1793–1830', *B. J. Hist. Sci.*, 13 (1980) 27–43 (34).

38. M. P. Crosland, *The Society of Arcueil: A View of French Science at The Time of Napoleon I* (Heinemann, 1967).

39. M. P. Crosland, *Gay-Lussac: Scientist and Bourgeois* (Cambridge: Cambridge University Press, 1978) p. 4.

40. M. P. Crosland, *The Society of Arcueil*, pp. 90–1; B. Haines, 'The Inter-relations between Social, Biological, and Medical Thought, 1750–1850: St Simon and Comte', *B. J. Hist. Sci.*, 11 (1978) 19–35 (21).

41. E. Frankel, 'Career-making in post-revolutionary France: the case of Jean-Baptiste Biot', *B. J. Hist Sci.*, 11 (1978) 36–48. Biot's case is exceptionally well documented.

42. Outram, 'Politics and Vocation: French Science', p. 33.

43. R. Fox, 'Scientific Enterprise and the Patronage of Research in France 1800–1870', *Minerva*, 11 (1973) 442–73 (446).

44. It must be said that Napoleon established the French beet-sugar industry, and in 1810 eased the burden of salt-tax (a grave impediment to the glass, soap and soda industries) in order to give 'new proofs of the interest we take in this kind of industry'. On the other hand his Continental System denied France the opportunity of importing sulphur from Sicily from 1807 to 1810 (J. G. Smith, *Heavy Chemical Industry in France*, p. 262).

45. R. R. Maras, 'Napoleon I and Education in the Sciences', *Actes du XIIe Congrès International d'Histoire des Sciences* (Paris, 1968, 1971), vol. XI, pp. 87–91.

46. L. Pearce Williams, 'Science, Education and Napoleon I', *Isis*, 47, (1956) 369–82.

47. Fox, 'Scientific Enterprise in France', p. 445.

48. R. Fox, *The Caloric Theory of Gases from Lavoisier to Regnault* (Oxford: Clarendon Press, 1971) pp. 231–2.

49. C. C. Gillispie, 'The Natural History of Industry', *Isis*, 48 (1957) 398–407; reprinted in A. E. Musson (ed.), *Science, Technology and Economic Growth in the Eighteenth Century* (Methuen, 1972) pp. 121–135.

50. Ibid., p. 129.

51. Ibid., p. 131.

52. C. C. Gillispie, 'The discovery of the Leblanc process', *Isis*, 48 (1957) 152–170. On Leblanc see R. Fox, *D.S.B.* and ref. 19.

53. J. Langins, 'The Decline of Chemistry'.

54. Cited in R. Grant, *History of Physical Astronomy* (1852) p. 108; the extreme left-wing views of Arago suggest he may not be well-categorised as a spokesman for Laplace and Berthollet.

55. M. P. Crosland, 'The first reception of Dalton's atomic theory in France', in D. S. L. Cardwell (ed.) *John Dalton and the Progress of Science* (Manchester: Manchester University Press, 1968) pp. 274–87.

56. R. Fox, 'The Rise and Fall of Laplacian Physics', *Hist. Stud. Phys. Sci.*, 4 (1974) 89–136 (136).

8. RADICAL SCIENCE IN BRITAIN, 1790–1830

1. Letter dated 1 November, 1792: Public Records Office, Home Office Papers, H.O., 42. 22. See C. Emsley, 'The Home Office and its sources of information and investigation 1791–1801', *Eng. Hist. Rev.*, 94 (1979) 532–61.

2. Letter dated 21 July, 1792: P.R.O., H.O., 42. 208; the letter went on to claim that the Vice-Chancellor, who had not been aware of Beddoes' political opinions when he recommended him to Lord Guildford, was now supporting Willoughby's views.

3. F. W. Gibbs and W. A. Smeaton, 'Thomas Beddoes at Oxford', *Ambix*, 9 (1961) 47–49; T. H. Levere, 'Dr Thomas Beddoes at Oxford: Radical Politics in 1788–1793 and the fate of the Regius Chair in Chemistry', Ibid., 28 (1981) 61–69.

4. T. H. Levere, 'Dr Thomas Beddoes and the Establishment of his Pneumatic Institution: A tale of Three Presidents', *Notes and Records*, 32 (1977) 41–49.

5. *Anti-Jacobin Review and Magazine,* 6 (1800) 111–18; 424–428.

6. T. Beddoes, *Notice of Some Observations Made at the Medical Pneumatic Institution* (Bristol: 1799).

7. T. M. Levere, 'Beddoes and his Pneumatic Institution', p. 47, remarks that this was legitimate since Beddoes associated social and material conditions with the causes of sickness.

8. See R. B. Rose, 'The Priestley riots of 1891', *Past and Present*, 18 (1960) 68–88.

9. Thus 'Manchester in the 1790s, with the French Revolution going sour after its early idealism, was an uncomfortable place for Radicals' (K. R. and W. V. Farrar and E. L. Scott, *Royal Institute of Chemistry Reviews,* 4 (1971) 42).

10. I. Inkster, 'The Social Context of an Educational Movement: A Revisionist Approach to the English Mechanics' Institutes, 1820–1850', *Oxford Rev. Educ.*, 2 (1976) 277–307 (291).

11. Ibid., pp. 291, 302–3.

12. Ibid., pp. 304–5.

13. For example, I. Inkster, 'Science and society in the metropolis: a preliminary examination of the social and institutional context of the Askesian Society of London, 1796–1807', *Ann. Sci.*, 34 (1977) 1–32; J. W. S. Cassels, 'The Spitalfields Mathematical Society', *Bulletin London Mathematical Society*, 11 (1979) 241–258; J. N. Hays, 'Science in the City: The London Institution, 1819–40', *B. J. Hist. Sci.*, 7 (1974) 146–162.

14. W. H. Brock, 'The London Chemical Society 1824', *Ambix*, 14 (1967) 133–9; C. A. Russell, N. G. Coley and G. K. Roberts, *Chemists by Profession* (Milton Keynes: Open University Press, 1977) pp. 56, 58–61, 65, 72.

15. First issue 13 March 1824; last issue 16 April 1825; 2 vols.

16. *The Chemist*, 1 (1824) 221.

17. Ibid., p. 167.

18. Ibid., p. 405.

19. Ibid., 2, 162–7.

20. Ibid., p. 168.

21. W. H. Brock, 'The London Chemical Society', p. 138.

22. The collapse of the Chemical Society may well be connected with the failure to attract (and pay) lecturers of sufficient ability as the correspondent suggested. But this, and the demise of *The Chemist*, illustrates the difficulty of sustained progress without the support of the science establishment, as well as the intense competition for patronage by a large number of similar organisations and periodicals. Apart from a burgeoning number of London institutions for popular science one observer noted 'the increase of periodical publications is one of the intellectual phenomena of the day' (*Panoramic Miscellany*, 1 (31 Jan. 1826) 1).

23. *The Chemist*, 2 (1824) 168.

24. Pigot's *Directory*, (1827).

25. W. H. Brock, 'The London Chemical Society', p. 137; *The Chemist* perpetrated several misprints of initials.

26. *Qu. J. Chem. Soc.*, 8 (1856) 109–10.

27. At 8 Austin Friars Passage (Post Office Directories, 1821–1824), and at 26 New Broad Street (PO Directories 1825–1830).

28. J. F. Harrison was a Wine Merchant with an office at the same address as Marreco in New Broad Street. His brother, William Harrison, of Whitburn near Sunderland, became Managing Director of the Stanhope and Tyne Railway, and founder of one of the greatest engineering families in the North East, see W. W. Tomlinson, *The North Eastern Railway, its Rise and Development* (Longmans, 1914) p. 214.

29. Probably members of the Sandemanian family into which Michael Faraday married. From 1832 the firm of Harrison, Barnard & Co., Wine Merchants, also appears at 26 New Broad Street (Robson's *Directory*, 1832–1839).

30. Tomlinson, *The North Eastern Railway*, p. 429.

31. Minutes of Durham Junction Railway, Public Records Office, British Transport Commission Archives, D.J.R. 1/1.

32. Letters from Marreco to John Buddle 7 Feb. to 21 Nov. 1842 in Buddle Papers at Durham Record Office, N.C.B./J.B., items 950–955.

33. Naturalisation Certificate for Walter Freire-Marreco, Jan. 1878, (P.R.O., H.O. 45/9452, 70135); 1871 census for 18 Saville Row, Newcastle (R. G. 10/5089) identifies Walter as a son of Marreco's widow.

34. At 3 St Thomas' Square, Newcastle (Ward's *Directory*, 1853).

35. G. B. Hodgson, *The Borough of South Shields* (South Shields: 1903) p. 402.

36. Baptismal Record of daughter Laura, 9 Feb. 1838, Tynemouth Parish Registers; he is described as 'merchant of Dockwray Square', a residential area for solicitors, physicians, ship-owners etc.

37. Accounts of Durham Junction Railway, P.R.O., B.T.C./D.J.R./1/1; he held £250 in shares in 1834, £5000 in 1842.

38. His wife was born in about 1806 (1871 census), but he may have been considerably older.

39. *PO Directory*, 1826.

40. W. H. Brock, 'The London Chemical Society', p. 137, makes this definite identification.

41. *The Chemist*, 1 (1824) 206; letter dated 28 May.

42. *D.N.B.*, 19, 35.

43. *The Chemist*, 2 (1824) 80, 96.

44. Pigot's *Directories*, 1825–1827.

45. *The Chemist*, 2 (1824) 162.

46. Pigot's *Directories*, 1826–7; 1828–9; Robson's *Directory*, 1830.

47. British Library, Add. MS 27791, ff 184–222. Reprinted in D. J. Rowe (ed.), *London Radicalism 1830–1843, a selection from the papers of Francis Place* (London Record Society Publications, 1970) pp. 68–9.

48. British Library, Add. MS 27796, Ff. 296; reprinted in Rowe, *London Radicalism*, p. 120.

49. *D.N.B.*, and references cited therein.

50. This is the view of W. H. Brock, 'The London Chemical Society', p. 134.

51. *The Chemist*, 1 (1824) 237.

52. Ibid., 1 (1824) vii.

53. Ibid., 1 (1824) 47; the statement is of course arrant nonsense but signifies a belief in radical science, uncluttered by the extravagant apparatus available only to the

rich and, in any case, unnecessary. If this interpretation is correct it was a very early expression of the belief that a political view could affect scientific research as opposed to scientific communication.

54. Ibid., 2 (1824) 47; Somerset House at that time was the home of the Royal Society whose method of rejecting candidates was known as 'black-balling'; the Royal Institution was situated in Albemarle Street; the Athenaeum, founded that same year, was in part the creation of Davy (F. W. Gibbs, 'Davy, the Athenaeum and the Zoo', *J. Roy. Inst. Chem.*, 84 (1960), 2–4).

55. H. Brougham, *Practical Observations upon the Education of the People*, 6th edn (London, 1825) p. 3.

56. British Library, Francis Place papers, Add. MS 27823, ff. 244.

57. *Mechanics' Magazine* (11 December 1824) p. 190.

58. British Library, Francis Place papers, Add. MS 27823, ff. 244, 369.

59. *Mechanics' Magazine* (11 December 1824) p. 191.

60. Several of these are noted by P. Weindling, 'Science and Sedition: how effective were the Acts Licensing Lectures and Meetings 1795–1819?', *B.J.Hist.Sci.*, 13 (1980) 139–153 (150).

61. D. M. Knight, 'Chemistry in palaeontology: the work of James Parkinson (1755–1824), *Ambix*, 21 (1974) 78–85; W. H. McMenemy, 'James Parkinson, 1755–1824; a biographical essay', in M. Critchley (ed.), *James Parkinson* (Macmillan, 1955) pp. 1–43.

62. D. Rockey, 'John Thelwall and the Origins of British Speech Therapy', *Med.Hist.*, 23 (1979) 156–75.

63. A. Goodwin, *The friends of liberty: the English Democratic Movement in the age of the French Revolution* (Hutchinson, 1979) pp. 333, 342, 358.

64. These are extensively discussed in the context of London science in I. Inkster 'London Science and the Seditious Meetings Act of 1817', *B.J.Hist. Sci.*, 12 (1979) 192–196 (193).

65. T. M. Parssinen, 'The Revolutionary Party in London, 1816–1820', *Bull. Inst. Hist. Res.*, 45 (1972) 266–82.

66. Inkster, 'London Science', p. 193.

67. Weindling, 'Science and Sedition', p. 146; he has examined the Middlesex Licensing records at the Greater London Record Office: in at least 37 other Record Offices negative results were obtained in searches for further licences.

68. One institution for which a licence was refused in 1817 was the City Philosophical Society, probably on the grounds that it dealt with political as well as scientific issues (Weindling, 'Science and Sedition', p. 145). It was founded by John Tatum in 1808, and attended by young Michael Faraday from 1810. As a fledgling lecturer he addressed the Society from 1816 to 1818 (L. P. Williams, *Michael Faraday* (Chapman and Hall, 1965) pp. 15–18).

69. Inkster, 'London Science', p. 195.

70. Weindling, 'Science and Sedition', p. 151.

71. N. G. Coley, 'The Animal Chemistry Club; Assistant Society to the Royal Society', *Notes and Records*, 22 (1967) 173–185.

72. Weindling, 'Science and Sedition', p. 149. But see also I. Inkster 'Seditious Science: a reply to Paul Weindling', *B.J.Hist.Sci.*, 14 (1981) 181–7.

73. T. H. Levere, 'Beddoes and his Pneumatic Institution', p. 44.

74. H. Davy, 'A Discourse Introductory to a Course of Lectures on Chemistry', 1802, in John Davy (ed.), *The Collected Works of Sir Humphry Davy* (1839) vol. II.

75. J. Z. Fullmer, 'Humphry Davy, Reformer', in S. Forgan (ed.), *Science and the Sons of Genius: Studies on Humphry Davy* (Science Reviews Ltd., 1980) pp. 59–94.

76. A. C. Todd, *Beyond the Blaze: a Biography of Davies Gilbert* (Truro: D. Bradford Barton, 1967).

77. J. B. Morrell, 'Professors Robison and Playfair, and the *Theophobia Gallica*: Natural Philosophy, Religion and Politics in Edinburgh, 1789–1815', *Notes and Records*, 26 (1971) 43–63 (49).

78. J. Priestley, *A Letter to . . . Edmund Burke* (Birmingham, 1791) pp. 139–43.

9. SCIENCE FOR THE MASSES, 1825–1850

1. On the origins of the R.I. see M. Berman, *Social Change and Scientific Organisation. The Royal Institution 1799–1844.* (Heinemann, 1978); K. D. C. Vernon, 'The Foundation and Early Years of the Royal Institution', *Proc. Roy. Inst.*, 39 (1963) 364–402.

2. T. Webster, *Autobiography*, bound MS in Royal Institution Library (c. 1837), f. 13; see also A. D. R. Caroe, *The House of the Royal Institution* (R.I., 1963) pp. 21–2.

3. Webster, *Autobiography*, f. 14.

4. Webster, ibid., f. 15; this is presumably the basis of the rather garbled reference about 'the back door' being 'bricked up' in J. D. Bernal, *Science in History* (Penguin, 1969) vol. II, p. 540.

5. On Dawes see D. Layton, *Science for the People: The Origins of the School Science Curriculum in England* (Allen and Unwin, 1973); and 'Science in the Schools: The First Wave – A study of the Influence of Richard Dawes (1793–1867)', *B.J. Educ. Stud.*, 20 (1972) 38–57.

6. D. Thompson, 'Queenwood College, Hampshire', *Ann. Sci.*, 11 (1955) 246–254.

7. See R. D. Altick, *The English Common Reader: A social history of the mass reading public, 1800–1900* (University of Chicago Press, 1957).

8. J. R. Partington, *A History of Chemistry* (Macmillan, 1962) vol. III, pp. 707–8.

9. L. P. Williams, *Michael Faraday* (Chapman and Hall, 1965) p. 19.

10. They are singled out in E. Frankland, Autobiographical *Sketches from the Life of Sir Edward Frankland*, 2nd edn (1902) p. 8.

11. F. Darwin, *Charles Darwin: His Life told in an Autobiographical Chapter, and in the selected series of his published letters*, 2nd edn (John Murray, 1902) p. 11; on Parkes see Partington, *History of Chemistry*, p. 706.

12. Frankland, *Sketches*, p. 3.

13. A. R. Wallace, *My Life* (1905) vol. I, pp. 20, 22–3.

14. J. G. Crowther, *Scientific Types* (Cresset, 1968) p. 142.

15. H. Hartley, *Humphry Davy* (Nelson, 1966) p. 12.

16. E. A. Parnell, *The Life and Labours of John Mercer* (1886).

17. M. E. S. and K. R. S., *Sir Joseph Wilson Swan, F.R.S.* (Benn, 1929) p. 20.

18. See M. Moseley, *Irascible Genius: A Life of Charles Babbage, Inventor* (Hutchinson, 1964) p. 40.

19. J. A. Fleming, *Memories of a Scientific Life* (London and Edinburgh: Marshall, Morgan & Scott, 1934) p. 10–11.

20. A. Treneer, *The Mercurial Chemist: A Life of Sir Humphry Davy* (Methuen, 1963) p. 19.

21. H. Spencer, *An Autobiography* Williams and Norgate, 1904) vol. I, p. 87.

22. Frankland, *Sketches*, pp. 8, 53.

23. G. Birkbeck, *Mechanics' Magazine*, 1 (1823) 179.

24. T. Kelly, *George Birkbeck, Pioneer of Adult Education* (Liverpool: Liverpool University Press, 1957) pp. 31, 33.

25. Ibid., p. 74.

26. Ibid., p. 71.

27. Ibid., p. 75.

28. See H. Smith, *'The Society for the Diffusion of Useful Knowledge, 1826–1846: a social and bibliographical evaluation'*, (Halifax, Nova Scotia: Dalhousie University Occasional Papers no. 8, 1974).

29. Ibid., p. 34.

30. Amongst the plethora of literature one may specially mention Kelly, *Birkbeck*; T. Kelly, *A History of Adult Education in Great Britain*, 2nd edn (Liverpool: Liverpool University Press, 1970); D. S. L. Cardwell, *The Organisation of Science in England*, 2nd edn (Heinemann, 1972) pp. 39–44, 71–75; J. F. C. Harrison, *Learning and Living, 1790–1960* (Routledge and Kegan Paul, 1961) pp. 57–89; M. D. Stevens and G. W. Roderick, 'Science, the Working Classes and Mechanics' Institutes,' *Ann. Sci.*, 29 (1972) 349–60; James Hole, *An Essay on the History and Management of Literary, Scientific and Mechanics' Institutions* (1853).

31. For Example, M. Tylecote, *The Mechanics' Institutes of Lancashire and Yorkshire before 1851* (Manchester: Manchester University Press, 1957); I. Inkster, 'Science and the Mechanics' Institutes, 1820–1850: the case of Sheffield', *Ann. Sci.*, 32 (1975) 451–474; D. S. L. Cardwell (ed.), *Artisan to Graduate, Essays to Commemorate the Foundation in 1824 of The Manchester Mechanics' Institution* . . . (Manchester: Manchester University Press, 1974); Kelly, *Birkbeck*, (on the London Mechanics' Institution); S. Shapin, 'The Pottery Philosophical Society, 1819–1835: An Examination of the Cultural Uses of Provincial Science', *Sci. Stud.*, 2 (1972) 311–336 (includes a good short account of the Pottery Mechanics' Institute); C. Delisle Burns, *A Short History of Birkbeck College* (University of London Press, 1924); L. J. Dyer, 'Newcastle Mechanics' Institute', *Adult Education*, 22 (1949–50) 122–9, 205–12; Kelly *Birkbeck*, gives full bibliography up to 1957.

32. Kelly, *Birkbeck*, p. 225.

33. Kelly, *Birkbeck*, pp. 332–339.

34. A. R. Stanley, *The Life and Correspondence of Thomas Arnold D.D.* (1891), p. 225 (letter 2 July 1834).

35. F. Engels, 'The Condition of the Working Class in England in 1844', in *Karl Marx and Frederick Engels on Britain*, 2nd edn (Moscow: Foreign Languages Publishing House) (1962) p. 274–5.

36. Inkster, *'Science and The Mechanics' Institutes'*, p. 472; he points out the widespread if uncritical reproduction of Engels' views.

37. Henry Solly, *These Eighty Years*, (1893) vol. II, p. 250, cited in Altick, *The English Common Reader*, p. 192.

38. Tylecote, *Mechanics' Institutes*, p. 135.

39. For example, E. Royle, 'Mechanics' Institutes and the Working Classes, 1840–1860', *Hist. J.*, 14 (1971) 305–321.

40. Royle, ibid., 311.

41. I. Inkster, 'The Social Context of an Educational Movement: A Revisionist Approach to the English Mechanics' Institutes, 1820–1850', *Oxford Rev. Educ.*, 2 (1976) 277–307.

42. E. P. Thompson, *The Making of the English Working Class* (Penguin, 1968) p. 819.

43. R. G. Kirby, 'An Early Experiment in Workers' Self Education', in D. S. L. Cardwell (ed.), *Artisan to Graduate*, pp. 87–98; Kelly, *Birkbeck*, p. 74.

44. British Library, Francis Place Correspondence, Add. MS 27823 f. 251.

45. Kelly, *Birkbeck*, p. 86n.

46. The changed definition of 'mechanics', as well as the undoubted shift in social composition of the Preston Institution are brought out by these figures for membership in 1841/2: professional men 23.3 per cent, clerks and shopmen 20.6 per cent, tradesmen 18.4 per cent, manufacturers 9.7 per cent, joiners and operatives 8.3 per cent, mechanics 4.1 per cent, gentlemen 3.4 per cent, ladies 1.5 per cent, youths at

school 1.5 per cent, factory hands 1.5 per cent, bankers 0.7 per cent, miscellaneous 7.0 per cent. G. Timmins, D. Foster and H. Law, *Preston Polytechnic: The emergence of an Institution 1828–1978* (Preston: Preston Polytechnic, 1978) p. 11. In the earlier sense of the word 'mechanics' would include also factory hands, joiners and operatives, and probably some persons from the manufacturing industries, tradesmen, clerks and shopmen.

47. D. Hinton, 'Popular Science in England 1830–1870' (unpublished Ph.D thesis, Bath, 1980) has these figures for the percentage of science books in the Mechanics' Institute Libraries: Keighley 55.2 (1828), 29.8 (1847); Newcastle 36.7 (1825), 24.0 (1836), 20.1 (1860); Birmingham 27.5 (1828), 26.4 (1834), 18.0 (1839); Warrington 33.6 (1828), 20.2 (1836), 23.8 (1847), 19.0 (1859); Ashton-under-Lyme 25.2 (1849), 22.8 (1853).

48. Royle, 'Mechanics' Institutes and the Working Classes', p. 309.

49. Burns, *Birkbeck College*, p. 47.

50. J. L. and B. Hammond, *The Bleak Age*, rev. edn (Penguin, 1947) p. 162.

51. Altick, *The English Common Reader*, p. 198; there were of course many other lending agencies in Britain at that time.

52. H. Perkin, *The Origins of Modern English Society 1780–1880* (Routledge and Kegan Paul, 1969) p. 306.

53. See, for example, C. A. Russell, 'Edward Frankland and the Cheapside Chemists of Lancaster: an Early Victorian Pharmaceutical Apprenticeship', *Ann. Sci.*, 35 (1978) 253–273.

54. M. Tylecote, *Mechanics' Institutes*, p. 26.

55. A. Highmore, *Philanthropia Metropolitana: A view of the charitable institutions established in and near London* (1822) p. 185 (in connection with the London School for Instruction and Industry).

56. G. Birkbeck, *Mechanics' Magazine*, 1 (1823) 116.

57. Cited in Kelly, *Birkbeck*, p. 85. (From *Cobbett's Weekly Register* 15 November 1823).

58. H. Brougham, *The Objects, Advantages and Pleasures of Science* (1826) p. 41.

59. G. Birkbeck, *Mechanics' Magazine*, 1 (1823) 117.

60. Thus it is interesting that in a sample of 25 library lists examined by D. Hinton ('Popular Science in England') the one with the highest proportion of science books (55.2 per cent) belonged to the Keighley Institute which was founded by four working men (Kelly, *Birkbeck*, p. 215).

61. S. Shapin and B. Barnes, 'Science, Nature and Control: interpreting Mechanics' Institutes', *Soc. Stud. Sci.*, 7 (1977) 31–74 (67n).

62. Thus Frankland, appointed to his first post in London in 1845, observes 'the desultory experiments and reading in Lancaster, though useful and the best obtainable under the circumstances, really left me absolutely ignorant of chemical analysis – the foundation of chemical science. I was in fact astonished to find how little I knew, or, rather, how absolutely ignorant I was, and I entered upon a course of systematic analysis with great zeal and delight'. (*Sketches*, pp. 63–4); when he went back to Lancaster the following year, his first lecture was on chemical analysis, and a laboratory class was formed for the study of qualitative analysis (MS Records of Class Meeting of Lancaster Mechanics' Institute, 12 August 1846, in Lancaster Public Library).

63. J. B. Morrell, 'The Chemist Breeders: the Research Schools of Liebig and Thomas Thomson', *Ambix*, 19 (1972) 1–46 (43).

64. For example, the picture of the laboratory at the London Mechanics' Institution in *Mechanics' Magazine* (1828) reproduced in Kelly, *Birkbeck*, opp. p. 115.

65. E. Frankland, *Sketches*, pp. 61–2; inevitably, perhaps, the practical work 'was of a rather desultory and unorganised character'.

66. Kelly, *Birkbeck*, p. 100

67. K. R. Farrar, 'Dalton's scientific apparatus', in D. S. L. Cardwell (ed.), *John Dalton and the Progress of Science*, (Manchester: Manchester University Press, 1968) pp. 159–186.

68. Letter to Griffin, 14 April 1838, in Wellcome Historical Medical Library, London.

69. L. M. in *Mechanics' Magazine*, 1 (1823) 234.

70. J. S. Taylor, ibid., p. 184.

71. A Country Gentleman, *The Consequences of a Scientific Education to the Working Classes of this Country pointed out, and the Theories of Mr Brougham on that Subject confuted; in a letter to the Marquis of Landsdown*, (1826); the identity of the author has never been established.

72. *St James' Chronicle*, May 1825, cited in Burns, *Birkbeck College*, p. 28.

73. For an admirably lucid account see Shapin and Barnes, 'Science, Nature and Control' and references therein.

74. See, for example, R. Johnson, 'Educational Policy and Social Control in early Victorian England', *Past and Present*, no. 49 (1970) 96–119; L. Stone, 'Literacy and Education in England, 1640–1900', ibid., no. 42 (1969) 69–139.

75. A. Thomson in *The Scotsman*, 8 June 1825, cited in Shapin and Barnes, 'Science, Nature and Control', p. 35–6.

76. For example, in Birkbeck's opening address at the London Mechanics' Institution, where he quoted from his Baptist friend the Reverend Robert Hall to the effect that by knowledge 'we become less dependent for satisfaction upon the sensitive appetites', (Kelly, *Birkbeck*, p. 91).

77. I. Inkster, 'English Mechanics' Institutes', p. 280.

78. T. R. Jolly, 'The Origins and Work of the Institute', in *34th Annual Report of the Harris Institute, Preston* (Preston: 1917) pp. 19–57.

79. Letter to *Pottery Mercury*, 28 September 1825, cited by Shapin, 'Pottery Philosophical Society', p. 333.

80. House of Commons, Report from Select Committee on the State of Education (1834), p. 49, cited by Shapin and Barnes, 'Science, Nature and Control', p. 39.

81. J. Foster, *Essay on the Evils of Popular Ignorance* (1820).

82. Such was the case at Lancaster, and many other smaller centres.

83. Though Shapin and Barnes ('Science, Nature and Control', p. 55), assert that 'only in the coercive context of schools for children did blatantly teleological interpretations of nature survive for any length of time', the evidence is not stated.

84. H. Brougham, *A Discourse on Natural Theology*, 3rd ed (1835) p. 199.

85. J. L. and B. Hammond, *The Town Labourer, 1760–1832* (Longmans, 1966) especially pp. 258–273.

86. For example, Thomas Dick in his *On the Mental Illumination and Moral Improvement of Mankind* (Glasgow: 1835).

87. It is interesting to note that the London Mechanics' Institution, though anxious to let its theatre to outside organisations, declined in 1825 to entertain the so called 'Christian Evidence Society' – a deistic organisation whose commitment to natural theology was total (Kelly, *Birkbeck*, p. 121).

88. Shapin and Barnes, 'Science, Nature and Control', p. 56.

89. This has been suggested, for example by Thackray, in the different social context of the Lit. and Phil. movement: A. Thackray, 'Natural knowledge in cultural context'; the Manchester model', *Amer. Hist. Rev.*, 79 (1974) 672–709 (693).

90. The difficulty of providing such unequivocal evidence does not relieve historians of the responsibility of attempting to provide it; otherwise claims about 'intention' and what people 'really' meant would stand simply on the basis of some

metaphysical presuppositions. By the same token their opposites can be asserted with equal strength and conviction, but again evidence is necessary.

91. Shapin and Barnes, 'Science, Nature and Control', p. 68.

92. Hinton, 'Popular Science in England', pp. 239–48.

93. Shapin and Barnes., p. 50.

94. A point that is still relevant today, as anyone will know who has taken evening classes in science for technicians. On the 'tyranny of abstraction' in the schools context see D. Layton, *Science for the People*, pp. 166–79.

95. Inkster, 'Science and the Mechanics' Institute', p. 464.

96. D. Burns, *Mechanics' Institutions: Their Objects and Tendencies* (Glasgow: 1837) pp. 56–7; cited in Shapin and Barnes, 'Science, Nature and Control', p. 57.

97. For example, M. Tylecote, 'The Manchester Mechanics' Institution 1824–50', in D. S. L. Cardwell, (ed.), *Artisan to Graduate*, p. 66.

98. *The Scotsman*, 7 September 1872, cited in Shapin and Barnes, 'Science, Nature and Control', p. 67.

99. An argument advanced by Shapin and Barnes, ('Science, Nature and Control', p. 68) for their social control model relates to the popular scientific organisation founded in Philadelphia in 1824 known as the Franklin Institute. In this case, unlike most of the institutes in Britain, research flourished. And here alone the projectors were not preoccupied with controlling the masses. The implication is that a British concern to do so led to a failure to translate utilitarian rhetoric into research practice. Setting aside the propriety of comparing institutions in different countries, we can note that the Franklin Institute was different in other respects also; it was open to a much wider social range of members, was generously supported, and in 1830 received congressional appropriation to defray the cost of research into steam boilers (S. L. Wright, *The Story of the Franklin Institute* (Philadelphia: Franklin Institute 1938) pp. 29–31). See also B. Sinclair, *Philadelphia's Philosopher Mechanics: A History of the Franklin Institute 1824–1865* (Baltimore: Johns Hopkins University Press, 1974).

100. From an Address delivered 21 September 1858 at the unveiling of Isaac Newton's monument, Grantham, in Brougham's *Tracts, Mathematical and Physical* (1860) pp. 280–2.

101. H. Brougham, *Dialogue on Instinct* (Philadelphia: 1845) p. 6.

102. Having delivered his lecture on chemical analysis at the Lancaster Mechanics' Institution, the following week Frankland 'read a paper on the early history and progress of chemistry' (papers of Lancaster Mechanics' Institute, in Lancaster Public Library; report for 19 August 1846).

103. D. S. L. Cardwell, *Artisan to Graduate*.

104. Inkster, 'Science and the Mechanics' Institutes'.

105. Ibid., pp. 454–5.

106. Shapin, 'Pottery Philosophical Society'.

107. *Lancaster Guardian*, 1 Jan. 1910.

108. W. B. Stephens, *Adult Education and Society in an Industrial Town: Warrington 1800–1900* (Exeter: University of Exeter, 1980), pp. 46–97 (53).

109. Ibid., pp. 52–3.

110. Inkster, 'English Mechanics' Institutes', and 'Science and the Mechanics' Institutes'.

111. N. Longmate, *The Water Drinkers, A History of Temperance* (Hamilton, 1968), p. 40.

112. Anonymous correspondent in *Manchester Isis*, 11 Oct. 1823, possibly G. W. Wood, a leading merchant and later MP and activist in local scientific institutions (R. H. Kargon, *Science in Victorian Manchester* (Manchester: Manchester University Press, 1977) pp. 18–19).

113. This happened even in the most remote areas of Britain. For example in

about 1840 a course of lectures on mechanics was begun at Easdale on the Hebridean Isle of Seil by John Whyte, engineer at the slate works. With the gift of an air-pump from the local laird, the 'infant institution' became apparently the first to use Gaelic as a medium of instruction. Although slate mining was the only non-agricultural industry in the area (employing about ten per cent of the population), the parish appears to have been a place of bustling activity with few orgies of social unrest and perhaps four per cent of the population on the poor-roll. The mechanics' lectures are more likely to have been stimulated by mechanisation of the slate-mine and general philanthropy, Whyte being well-known both for his mechanical expertise and his Eldership of the Church. (*The New Statistical Account of Scotland* (Edinburgh and London, 1845) vol. VII, 'Argyle', pp. 71–83.)

10. STRONGHOLDS OF AMATEUR SCIENCE IN ENGLAND

1. C. Babbage, *Reflections on the Decline of Science in England* (1830; 1969 Gregg reprint), pp. 10–11.

2. J. Carrière, *Berzelius und Liebig, ihre Briefe von 1831–1845* (Munich and Leipzig, 1893) p. 134.

3. G. Kitteringham, 'Science in Provincial Society: The case of Liverpool in the Early Nineteenth Century', *Ann. Sci.*, 39 (1982), 329–348.

4. E. Kitson Clark, *The History of 100 Years of Life of the Leeds Philosophical and Literary Society* (Leeds: Jowett and Sowry, 1924).

5. W. A. Beanland, *The History of the Royal Institution of South Wales, Swansea, for the First Few Years Known as the Swansea Philosophical and Literary Society, 1835–1935* (Swansea: Royal Institution, 1935), p. 14.

6. T. Kelly, *George Birkbeck* (Liverpool: Liverpool University Press, 1957), pp. 210–12, 232.

7. Minute Book of Reading Room Society, 8 February 1826 and 21 April 1828 in Local Collection, Bedford Central Library.

8. *Our Columns, Journal of Bedford Literary and Scientific Institute and General Library* (Jan. 1892), p. 54–9; see also A. E. Baker, *The Library Story* (Library Association Dissertation, 1956).

9. Beanland, *Royal Institution of South Wales*, pp. 14–18.

10. *Sunderland Herald,* 26 September 1835, 22 October 1836, 14 January 1837, 14 October 1837, 19 October 1838.

11. S. Shapin, 'The Pottery Philosophical Society, 1819–1835; An Examination of the Cultural Uses of Provincial Science', *Science Studies,* 2 (1972) 311–336.

12. R. H. Kargon, *Science in Victorian Manchester* (Manchester: Manchester University Press, 1977) pp. 5–14, 41–49; H. J. Fleure, 'The Manchester Literary and Philosophical Society', *Endeavour*, 6 (1947) 147–151.

13. R. S. Watson, *The History of the Literary and Philosophical Society of Newcastle-upon-Tyne (1793–1896)* (1897) pp. 165, 167.

14. It had a mortgage of £6200 and a current account overdrawn by £389 (Watson, *The Lit. and Phil. of Newcastle-upon-Tyne*, p. 126).

15. J. W. Hudson, *The History of Adult Education* (1851), p. 167.

16. Watson, p. 169.

17. Thus Watson (*The Lit. and Phil. of Newcastle-upon-Tyne*, pp. 305–6) observes that there had long been 'considerable friction between members of this Society who cared most for the Library and those members who cared most for the Museum'.

18. See D. E. Allen, *The Naturalist in Britain* (Pelican, 1978).

19. Lynn Barber, *The Heyday of Natural History, 1820–1870* (Jonathan Cape, 1980) p. 23.

20. See also A. Clow, 'Fiscal Policy and the Development of Technology', *Ann. Sci.*, 10 (1954) 342–358 (356).

21. Allen, *Naturalist in Britain*, p. 123.

22. T. R. Goddard, *History of the Natural History Society of Northumberland, Durham and Newcastle-upon-Tyne, 1829–1929*, (Newcastle: Reid, 1929) pp. 1, 188–195.

23. Ibid., p. 30.

24. Ibid., pp. 67–8.

25. Kargon, *Science in Victorian Manchester*, pp. 14–16.

26. Presidential Address (Ralph Carr). *Trans. Tyneside Naturalists' Field Club* 1 (1846–50) 6.

27. W. B. Stephens, *Adult Education and Society in an Industrial Town: Warrington 1800–1900* (Exeter, University of Exeter, 1980) p. 40.

28. Allen, *Naturalist in Britain*, p. 164.

29. T. G. Elger, *Trans. Beds. Nat. Hist. Soc.*, 1 (1875–6) 29; the Society was formed after the lively correspondence in the *Gardeners' Chronicle* about one flower, the Sweet Flag, plentifully found on the banks of the Ouse. This was yet another Society whose energies expired after about ten years.

30. F. E. Kingsley, *Charles Kingsley: His Letters and Memories of His Life*, 3rd edn (1877) vol. II, p. 350. Kingsley's motives may be compared with those attributed to earlier promoters of both Lit. and Phil. Societies and Mechanics' Institutions. He saw natural science as an instrument for the democratic ideals of 'liberty, equality and brotherhood', and a 'Natural Science Club' as a co-operative venture for young people to gain access to expensive books and apparatus and as a 'great freemasonry which owns no difference of rank, creed, or of nationality', as witness the experience of Hugh Miller and Michael Faraday (C. Kingsley, *Scientific Lectures and Essays*, 2nd edn. (1890) pp. 13–18). See also W. A. Herdman, *Charles Kingsley and the Chester Naturalists* (Chester: Natural Science Soc., 1921).

31. Allen, *Naturalist in Britain*, p. 170.

32. Barber, *Heyday of Natural History*, p. 15.

33. The Centenary History of the British Association is O. J. R. Howarth, *The British Association for the Advancement of Science: A Retrospect 1831–1931* (BAAS, 1931); to celebrate its sesquicentenary two new books appeared; Roy MacLeod and Peter Collins, *The Parliament of Science, The British Association for the Advancement of Science, 1831–1981* (Science Reviews Ltd., 1981); A. Thackray and J. B. Morrell, *Gentlemen of Science: The Origins and Early Years of the British Association for the Advancement of Science* (Oxford: Oxford University Press, 1981). The last of these books was not available during the preparation of this chapter.

34. *Edinburgh Review*, 60 (1834–5) 377.

35. *Quarterly Review*, 43 (1831) 305–42.

36. Babbage, *The Decline of Science*, pp. viii and vi; Herschel's words were cited from an article on 'Sound' in the *Encyclopaedia Metropolitana*.

37. On this theme see also A. D. Orange, 'The Origins of the British Association for the Advancement of Science', *B. J. Hist. Sci.*, 6 (1972) 152–76.

38. C. Babbage, 'Account of the Great Congress of Philosophers at Berlin', *Edinburgh J. Sci.*, 10 (1829) 225–34; J. F. W. Johnston, 'Meeting of the Cultivators of Natural Science and Medicine at Hamburgh', ibid., n.s., 4 (1830/1) 189–244.

39. A. D. Orange, 'The Beginnings of the British Association, 1831–1851', in MacLeod and Collins, *The Parliament of Science*, pp. 43–64 (44).

40. The choice of York may have been connected with the fact that Brewster was a member of the Yorkshire Philosophical Society, or that the Society's President was the influential Vernon Harcourt, son of the Archbishop, or even that York was geographically convenient, being almost exactly half-way on the road between the two capitals London and Edinburgh.

41. For example, L. P. Williams, 'The Royal Society and the Founding of the British Association for the Advancement of Science', *Notes and Records*, 16 (1961) 221–33.

42. R. MacLeod, 'Retrospect: The British Association and its Historians', in MacLeod and Collins, *The Parliament of Science*, pp. 1–16, (7, 8).

43. Howarth, *The British Association*, p. 23.

44. Ibid., p. 27.

45. C.f. M. Berman, ' "Hegemony" and the Amateur Tradition in British Science', *J. Soc. Hist.*, 1 (1975) 30–50 (37).

46. J. Henry, Letter to Charles Coquerel, 1 October 1837, in N. Reingold (ed)., *The Papers of Joseph Henry* (Washington: Smithsonian Institution Press, 1979) vol. III, p. 511.

47. S. F. Cannon, *Science in Culture: the Early Victorian Period* (New York: Dawson and Science History Publications, 1978) p. 83.

48. A. D. Orange, in MacLeod and Collins, *The Parliament of Science*, p. 52.

49. J. Liebig, Letter to M. Faraday, 19 December 1844, in L. P. Williams (ed.), *The Selected Correspondence of Michael Faraday* vol. I, (1812–48) (Cambridge: Cambridge University Press, 1971) p. 430.

50. W. H. Brock, 'Advancing Science: the British Association and the Professional Practice of Science', in MacLeod and Collins, *The Parliament of Science*, pp. 89–117, (108).

51. Philip Lowe, 'The British Association and the Provincial Public', in MacLeod and Collins, *The Parliament of Science*, pp. 118–144 (135).

52. Lowe, ibid., p. 122.

11. THE RISE OF THE SPECIALIST

1. Letter to W. Whewell, 20 May 1840, in Trinity College Library, Cambridge, reproduced in L. P. Williams, (ed.), *The Selected Correspondence of Michael Faraday* (Cambridge: Cambridge University Press, 1971) vol. I (1812–1848), pp. 377–8.

2. W. Whewell, *The Philosophy of the Inductive Sciences, Founded upon Their History* (1840) p. 71.

3. Ibid., p. 113. Possibly at the 1834 BAAS meeting.

4. D. E. Allen, 'The Women Members of the Botanical Society of London 1836–1856', *B.J.Hist.Sci.*, 13 (1980) 240–54.

5. M. J. S. Rudwick, 'The Foundation of the Geological Society of London: its Scheme for Co-operative Research and its Struggle for Independence', *B.J.Hist. Sci.*, 1 (1963) 325–355 (348).

6. R. Porter, 'The Industrial Revolution and the Rise of the Science of Geology', in M. Teich and R. Young (eds), *Changing Perspectives in the History of Science: Essays in Honour of Joseph Needham* (Heinemann, 1973) 320–343 (325).

7. J. Phillips, *Memoirs of William Smith LL.D.*, (1844) p. 133.

8. Porter, 'The Rise of the Science of Geology', p. 323.

9. Ibid., p. 326.

10. Cited in his Presidential Address by Col. Portock, *Qu. J. Geol. Soc.*, 13 (1857) xxviii.

11. Broadly speaking controversy as to whether water or fire had been chiefly responsible for determining the shape of the earth's crust.

12. P. J. Weindling, 'Geological Controversy and its Historiography: The Prehistory of the Geological Society of London', in L. J. Jordanova and R. S. Porter (eds) *Images of the Earth* (Chalfont St Giles: British Society for the History of Science, 1979) pp. 248–271, (256).

13. M. Berman, *Social Change and Scientific Organisation: The Royal Institution, 1799–1844* (Heinemann, 1978) pp. 88–9.

14. Weindling, 'Geological Controversy' p. 254.

15. Berman, *Social Change and Scientific Organisation*, p. 91.

16. Rudwick, 'The Foundation of the Geological Society', p. 355.

17. Porter, 'The Rise of the Science of Geology', p. 327.

18. A. C. Todd, 'Origins of the Royal Geological Society of Cornwall', *Trans. Roy. Geol. Soc. Cornwall*, 19 (1959/60) 179–184.

19. J. Hutton, 'Theory of the Earth', *Trans. Roy. Soc. Edinburgh*, 1 (1795) 200.

20. Ramsay's father had been a chemical manufacturer who founded the Glasgow Chemical Society; after his death his widow took in lodgers, amongst whom was the young Lyon Playfair. Having caught the eye of the Prime Minister, Peel, for the excellence of his chemical talent, Playfair was appointed in 1842 as Chemist to the Geological Survey in Duke Street, Westminster; his assistants there included Edward Frankland and Hermann Kolbe. Meanwhile de la Beche had other irons in the fire. See P. J. McCartney, 'Henry de la Beche: *Observations on an Observer*', (Cardiff: Friends of National Museum of Wales, 1977).

21. See M. Reeks, *Register of the Associates and Old Students of the Royal School of Mines and History of the Royal School of Mines* (Royal School of Mines, 1920).

22. G. K. Roberts, 'The Establishment of the Royal College of Chemistry: an investigation of the social context of early-Victorian Chemistry', *Hist. Stud. Phys. Sci.*, 7 (1976) 437–485. See also C. A. Russell, 'Who is a Chemist?', in Russell, Coley, Roberts, *Chemists by Profession* (Milton Keynes: Open University Press, 1977) pp. 29–54; R. F. Bud, 'The discipline of chemistry: the origins and early years of the Chemical Society of London', (University of Pennsylvania, Ph.D. Dissertation 1980).

23. Roberts, 'The Establishment of the Royal College of Chemistry', p. 447.

24. At the very beginning of the century Rumford had sought to appeal to the agricultural interests of the great landowners in order to establish his Royal Institution. In this way Davy's early researches on agricultural chemistry and geology can be readily understood. (Berman, *Social Change and Scientific Organisation.*) Similar principles informed the establishment of many of the agricultural societies of Britain including the Royal Agricultural Society of England founded in 1838. See, for example, K. Hudson, *Patriotism with Profit: British Agricultural Societies in the Eighteenth and Nineteenth Centuries* (Hugh Evelyn, 1972).

25. Roberts, 'The Establishment of the Royal College of Chemistry', p. 456.

26. Letter from P. Pusey to W. Buckland, 27 October 1842, in Imperial College Archives, cited by G. K. Roberts, 'The Establishment of the Royal College of Chemistry', p. 457.

27. J. Thomson, *Report of Select Committee Appointed to Inquire into Manufactures, Commerce and Shipping* P.O., (1833) III, 224.

28. Russell, 'Who is a Chemist?' pp. 36–8.

29. Bud, 'The discipline of chemistry'; accounts of the Chemical Society's formation may also be found in T. S. Moore and J. C. Phillips, *The Chemical Society 1841–1941* (Chemical Society, 1947) pp. 11–20, and in a short article by Robert Warington Jnr. on 'The Foundation of the Chemical Society', in *The Jubilee of the Chemical Society of London* (The Chemical Society, 1896) pp. 115–122.

30. Speech by R. Warington Snr., in *The Jubilee of the Chemical Society*, p. 18.

31. J. B. Morrell, 'The Chemist Breeders: The Research Schools of Liebig and Thomas Thompson', *Ambix*, 19 (1972) 1–46.

32. W. Francis and H. Croft, 'Notices of the Labours of Continental chemists', *Phil. Mag.*, (3), 18 (1841) 202.

33. Speech by W. J. Russell, *The Jubilee of the Chemical Society*, p. 11.

34. Ibid., p. 118.

35. Ibid., pp. 116–17.

36. Roberts, 'The Establishment of the Royal College of Chemistry', p. 473.

37. See W. A. Campbell, 'The Newcastle Chemical Society and its Illustrious Child', *Chem. and Ind.*, (1968) 1463–6.

38. It was proudly claimed that only two other learned societies in the north of England published original research (*Trans. Newcastle Chem. Soc.*, 1 (1868–9) 95–6).

39. Ibid., 2 (1870–4) 106, 165.

40. Ibid., p. 172.

41. Essentially uninterested in making money, Glover emerged from the manual working class, and began life as a plumber's apprentice. His famous 'Glover Tower', though not patented, was reputed to have saved the country at least £300,000 per annum. (See D. W. F. Hardie, *Chem. Age,* 78 (1957) 816.) On the industrial action probably referred to see E. Allen, J. F. Clarke, N. McCord, and D. J. Rowe, *The North-East Engineers' Strikes of 1871* (Newcastle: Frank Graham, 1971).

42. *Trans. Newcastle Chem. Soc.*, 2, (1870–4), 174.

43. Ibid.

44. W. W. Proctor, President, Inaugural Address, 1873–4, *Trans. Tyne Chem. Soc.*, 3 (1873–4).

45. Ibid. (1876) Secretary's Report.

46. A detailed computerised analysis of the membership of these societies is currently proceeding.

47. See J. Hargreaves, *J. Soc. Chem. Ind.*, 40 (1921) 86R–87R.

48. S. Middlebrook, *The Advancement of Knowledge in Newcastle-upon-Tyne: The Literary and Philosophical Society as an Educational Pioneer* (Literary and Philosophical Society of Newcastle-upon-Tyne, 1974).

49. R. Fox, 'The *Savant* confronts his peers: Scientific Societies in France, 1815–1914', in R. Fox and G. Weisz (eds), *The Organisation of Science and Technology in France 1808–1914*, (Cambridge: Cambridge University Press, 1980) pp. 241–82.

50. Ibid., pp. 244–58.

51. Ibid., p. 257.

52. Ibid., p. 244.

53. Ibid., pp. 256–7.

54. Ibid., pp. 250–1.

55. P. Laffitte, 'La Société Chimique de France, 1857–1957', *Proc. Chem. Soc.,* 1957, pp. 172–74; *Centenaire de la Société Chimique de France (1857–1957)* (Paris, Masson, 1957).

56. W. Ruske, *100 Jahre Deutsche chemische Gesellschaft* (Verlag Chemic GmbH, 1967).

57. H. Skolnik and K. M. Reese, *A Century of Chemistry*, (Washington D.C.: American Chemical Society, 1976).

58. A. Gautier, in *Centenaire,* p. 11.

59. W. Ruske, p. 29.

60. H. Skolnik and K. M. Reese, p. 5.

61. A. Gautier, *Centenaire*, p. 7.

62. Ibid., p. 17.

63. *The Jubilee of the Chemical Society of London*, (London: Chem. Soc., 1896), p. 39.

64. e.g., his *History of the Conflict between Religion and Science*, (London, 1875).

65. W. Ruske, p. 38.

12. THE ROAD TO PROFESSIONALISATION

1. Letter dated 1799 in Rolleston Collection.

2. Letter dated 8th March 1801, in Anne Treneer, *The Mercurial Chemist*, (Methuen, 1963) p. 74.

3. A. Findlay, 'The Royal Institute and the Profession of Chemistry', *Proc. Inst. Chem.*, 68 (1944) 211–216 (211).

4. See E. Shils, 'The Profession of Science', *Adv. Sci.*, 24 (1968) 469–80; J. Ben-David, 'The Profession of Science and its Powers', *Minerva*, 10 (1972) 362–383; E. Mendelsohn, 'The Emergence of Science as a Profession in Nineteenth Century Europe', in K. Hill (ed.), *The Management of Scientists* (Boston USA: Beacon Press, 1964) pp. 3–48; D. M. Knight, 'Science and Professionalism in England, 1770–1830', *Proc. XIVth Int. Congress Hist. Sci.*, Tokyo 1 (1974) 53-67. In a paper on 'The Process of Professionalization in American Science: the Emergent Period, 1820–1860' *Isis*, 58 (1967) 151–66. G. H. Daniels appears to equate 'the professional' with 'the trained specialist'. He identifies four stages on the professionalisation process in the USA: (1) pre-emption, (2) institutionalisation, (3) legitimation, and (4) the attainment of professional autonomy. The model does not appear to fit British developments, as will be seen in the following Section. For the case of France, see Ch. 7, especially references 33 and 34.

5. C. Babbage, *Reflections on the Decline of Science in England* (1830) p. 10.

6. J. Z. Fullmer, 'Technology, Chemistry and the Law in Early Nineteenth Century England', *Technology and Culture*, 21 (1980) 1–28.

7. W. H. Brock, 'The Spectrum of Science Patronage', in G L'E Turner (ed.) *The Patronage of Science in the Nineteenth Century* (Leyden: Noordhoff International Publishing, 1976) pp. 173–206; thus A. W. Hofmann in 1853 said it was possible in this way to earn £8000–£9000 per annum (p. 186).

8. *Census returns for Great Britain for 1861* (P.P. 1863) LIII, 231.

9. Leone Levi, 'On the Progress of Learned Societies: Illustrative of the Advancement of Science in the United Kingdom during the last thirty years', *Rep. Brit. Assoc., Trans. Sections* (Norwich: 1868) 169–72.

10. D. S. L. Cardwell, *The Organisation of Science in England*, 2nd. edn (Heinemann, 1972) p. 100.

11. *English Mechanic* (1868) 318.

12. F. M. L. Thompson, *The Chartered Surveyors: The Growth of a Profession* (Routledge and Kegan Paul, 1968) p. 149.

13. Ibid., p. 65.

14. C. A. Russell, G. K. Roberts and N. G. Coley, *Chemists by Profession* (Milton Keynes: Open University Press, 1977).

15. This section is largely taken from the author's chapters in *Chemists by Profession*.

16. *Chem. News.*, 15 (1867) 163.

17. Ibid., 33 (1876) 220.

18. Ibid., p. 146 (report of discussion at Anniversary Meeting of the Chemical Society on 30 March 1876).

19. Ibid., pp. 246–7 (undated letter from N. S. Maskelyne).

20. S. A. Russell, 'The growing role of chemical analysis', in *Chemists by Profession*, pp. 94–112.

21. *Chem. News* 33 (1876) 27–9; C. R. A. Wright was then Lecturer in Chemistry at St Mary's Hospital, Paddington, (obituary in *Proc. Roy. Soc.* 57 (1894) v–vii).

22. *Chem. News*, 33 (1876) 73.

23. Ibid., p. 27

24. Ibid., p. 94 (letter from A. H. Allen, 21 Feb., 1876).

25. Ibid., p. 89 (editorial).

26. Ibid., p. 102.

27. J. C. Carr and W. Taplin, *History of the British Steel Industry* (Oxford. Blackwell, 1962) p. 218. The advice was proffered to the Directors of the West Cumberland

Steelworks concerning the appointment of a notable metallurgical chemist. G. J. Snelus.

28. *Chem. News*, 33 (1876) 107 (undated letter from Charles T. Kingzett).

29. C. A. Russell, 'The birth of the Institute', *Chemists by Profession*, pp. 135–157.

30. C. A. Russell, 'Edward Frankland – the first President' *Chem. Brit.* 13 (1977) 4–7.

31. John Lankford, 'Amateurs versus Professionals: the controversy over telescope size in late Victorian science. *Isis*, 72 (1981) 11–28. 'Professionalism' in this article is mainly used in the rather limited sense of full-time employment.

32. Cited ibid., p. 21.

33. Ibid., p. 22.

34. John Lankford 'Amateurs and Astrophysics: a neglected aspect in the development of a scientific specialty', *Soc. Stud. Sci.*, 11 (1981) 275–303.

35. W. Noble, *J. Brit. Astronom. Assoc.*, 1 (1890) 49–55, 'The field of research has become so illimitable that the individual observer must confine himself to a very circumscribed area of it if he is to do anything useful' (p. 50).

36. See C. A. Russell, *The History of Valency* (Leicester University Press, 1971).

37. J. G. O'Connor and A. J. Meadows, 'Specialization and Professionalization in British Geology', *Soc. Stud, Sci.*, 6 (1976) 77–89. The authors define a professional geologist as simply 'someone who obtains his livelihood primarily from geological work' (p. 78). Whether this is a 'reasonable definition nowadays' for a study that extends to the 1950s must be open to some doubt. See also Roy Porter 'Gentlemen and Geology: the emergence of a scientific career, 1660–1920', *Hist. J.*, 21 (1978) 809–36.

38. P. D. Lowe, 'Amateurs and professionals: the institutional emergence of British plant ecology', *J. Soc. Bibliography Nat. Hist.*, 7 (1976) 517–35.

39. A. G. Tansley, 'The Problems of Ecology', *New Phytologist*, 3 (1904) 191–200 (196).

40. R. Fox, 'The *Savant* confronts his peers: Scientific Societies in France, 1815–1914', in R. Fox and G. Weisz (eds.), *The Organisation of Science and Technology in France 1808–1914* (Cambridge: Cambridge University Press, 1980) pp. 241–82 (265).

41. Ibid., p. 258.

13. THE STATE AND SCIENCE

1. L. H. Brockway, *Science and Colonial Expansion: the role of the British Royal Botanic Gardens* (Academic Press, 1978) p. 186; Paxton gained unique success in growing the Amazonian water-lily, Victoria regia.

2. Charles Babbage dyspeptically describes them as 'miserable trees', several of which were 'on the verge of extinction'; see Babbage, *The Exposition of 1851; or views of the industry, the science, and the government of England*, (1851) p. 28. He was using the occasion to mount a further attack on the scientific establishment and the Government, and further to advertise his ingenious 'calculating engines'.

3. See J. G. Crowther, *Statesmen of Science* (Cresset Press, 1965) for Lyon Playfair (pp. 105–171), and the Prince Consort (pp. 175–209).

4. The whole structure was demolished in 1852 and re-erected (with some alteration), as the Crystal Palace, on high ground near Sydenham where it remained until destroyed by fire in 1936. Paxton himself was knighted and in 1854 became Liberal MP for Coventry.

5. See Babbage *The Exposition of 1851*, p. 189.

6. L. Playfair (letter to Lord Taunton), *J. Soc. Arts.*, 15 (1867) 477–8.

7. Peter Lundgreen, 'The Organisation of Science and Technology in France: a German perspective', in R. Fox and G. Weisz, *The Organisation of Science and Technology in France 1808–1914*, (Cambridge: Cambridge University Press, 1980) pp. 311–332.

8. M. Arnold, *Schools and Universities on the Continent* (1868) p. 232.

9. R. Fox, 'Scientific Enterprise and Patronage of Research in France, 1800–1870', *Minerva*, 11 (1973) 442–73.

10. On this, as on much else in this chapter, see D. S. L. Cardwell, *The Organisation of Science in England,* 2nd edn (Heinemann, 1972) especially pp. 111–186.

11. G. S. Emmerson, *John Scott Russell: a great Victorian engineer and Naval Architect* (John Murray, 1977) pp. 253–8.

12. On this see R. MacLeod 'The X-club, a social network of science in late Victorian England', *Notes and Records*, 24 (1970) 305–22; J. V. Jensen, 'The X-club: fraternity of Victorian scientists', *B. J. Hist. Sci.,* 5 (1970) 63–72; Ruth Barton, 'The X-club: Science, Religion and Social Change in Victorian England', (University of Pennsylvania Ph.D. dissertation, 1976).

13. On the seventh Duke see Crowther, *Statesmen of Science,* pp. 213–233.

14. For references on Strange see Crowther, *Statesmen of Science,* pp. 237–269.

15. See R. MacLeod, 'Resources of Science in Victorian England: the endowment of science movement, 1868–1900' in P. Mathias (ed.), *Science and Society, 1600–1900* (Cambridge: Cambridge University Press, 1972) pp. 111–166.

16. Barton, 'The X-club', p. 188.

17. *Rep. Brit. Assoc. Adv. Sci.*, 20 (1851) li.

18. *English Mechanic*, (22 July 1881) 472.

19. Ibid., (17 December 1880), 349. See also A. J. Meadows, *Science and Controversy: a biography of Sir Norman Lockyer* (Macmillan, 1972) p. 349.

20. *English Mechanic* (25 February 1881) 586–7.

21. R. MacLeod, 'The Ayrton Incident: a commentary on the relations of Science and Government in England, 1870–1873' in A. Thackray and E. Mendelsohn (eds), *Science and Values: Patterns of Tradition and Change* (New York: Humanities Press, 1974) pp. 45–77.

22. *Rep. Brit. Assoc. Adv. Sci.,* (1874) lxvi–xcvii.

23. J. Morley, *The Life of W. E. Gladstone* (1903) vol. I, p. 209.

24. MacLeod, 'Resources of Science' p. 141.

25. MacLeod, ibid., p. 145.

26. On these see W. H. Brock, 'The Spectrum of Science Patronage', in G. L'E. Turner (ed.) *The Patronage of Science in the Nineteenth Century* (Leyden: Noordhoff International Publishing, 1976) pp. 173–206 (178–186).

27. Brockway, *Science and Colonial Expansion*.

28. MacLeod, 'Resources of Science', pp. 165–6.

29. R. Fox, 'Scientific enterprise and patronage of research in France, 1800–1870'.

30. M. P. Crosland, 'Science and the Franco-Prussian War', *Soc. Stud. Sci.* 6 (1976), 185–214.

31. P. F. Frankland and G. C. Frankland, *Pasteur* (London: Cassell, 1898), pp. 108–9.

32. R. J. Dubois, *Louis Pasteur, Freelance of Science*, (London: Gollancz, 1951), p. 78.

33. Crosland, 'Science and the Franco-Prussian War'.

34. J. B. Dumas, *Comptes rendus*, 1871, *72*, 265–6.

35. H. St.-C. Deville, ibid., 1871, *72*, 237–8.

36. Crosland, 'Science and the Franco-Prussian War' p. 191.

37. J. Ben-David, *The Scientist's role in society: A comparative study*, (Prentice-Hall, 1971) p. 106.

38. M. J. Nye, 'The scientific periphery in France: The Faculty of Sciences at Toulouse (1880–1930)', *Minerva*, 13 (1975) 374–403).

39. G. Weisz, 'The French universities and education for the new professions, 1885–1914: an episode in French university reform', *Minerva*, 1979, *17*, 98–128.

40. e.g., Matthew Arnold, *Higher Schools and Universities in Germany*, Second edition (1892).

41. *Trans. Newcastle Chem. Soc.*, 1872, *2*, 175–6.

42. W. V. Farrar, 'Science and the German university system, 1790–1850', in M. P. Crosland (ed.), *The Emergence of Science in Western Europe*, (London: Macmillan, 1975), pp. 179–92 (189).

43. W. O. Henderson, *The State and the Industrial Revolution in Prussia, 1740–1870*, (Liverpool University Press, 1958).

44. F. Plesch, 'Scientific organisation and science policy in Imperial Germany, 1871–1914: the foundation of the Imperial Institute of Physics and Technology', *Minerva*, 8 (1970) 557–80.

45. Ibid., p. 561.

46. See also J. A. Joynson, 'The Chemical Reichsanstalt Association: Big Science in Imperial Germany', (Ph.D. thesis, Princeton, 1980), especially pp. 1–76 (University Microfilms Ltd., no. 808563).

47. D. Cahan, 'Werner Siemens and the origin of the Physikalich-Technische Reichsanstalt, 1872–1887', *Hist. Stud. Phys. Sci.*, 12 (1982), 253–83.

48. Ibid., p. 276.

49. Ibid., p. 254.

50. Ibid., p. 283.

51. Brock, 'The Spectrum of Science Patronage', p. 201.

52. G. Ehrhart, 'One hundred years of research', in E. Bäumler, *A Century of Chemistry*, (Dusseldorf: Econ Verlag, 1968), pp. 273–337.

53. F. Leprieur and P. Papon, 'Synthetic dyestuffs: the relations between academic chemistry and the chemical industry in nineteenth-century France', *Minerva*, 17 (1979) 197–224 (223).

54. Lundgreen, 'The Organisation of Science and Technology in France: a German Perspective', p. 332.

55. Ibid.

14. EPILOGUE: SCIENCE TRIUMPHANT?

1. On Dalton's funeral and its significance, see Elizabeth C. Patterson, *John Dalton and the Atomic Theory: the biography of a natural philosopher* (New York: Doubleday, 1970).

2. See J. R. Moore, 'Charles Darwin lies in Westminster Abbey', *Biol. J. Linnean Soc.*, 17 (1982) 97–113.

3. R. M. Young, 'Natural theology, Victorian periodicals and the fragmentation of a common context', in C. Chant and J. Fauvel (eds), *Darwin to Einstein: Historical Studies on Science and Belief* (London: Longman/Open University Press, 1980) pp. 69–107.

4. Ibid., p. 73.

5. See, for example, F. M. Turner, *Between Science and Religion: the Reaction to Scientific Naturalism in late Victorian England* (New Haven: Yale University Press, 1974) and the same author's 'The Victorian conflict between science and religion: a professional dimension', *Isis*, 69, (1978) 356–76.

6. D. Oldroyd, *Darwinian Impacts: an introduction to the Darwinian revolution* (Milton Keynes: Open University Press, 1980) p. 238.

7. R. Westfall, essay on 'Marxism and the history of science', *Isis*, 72 (1981) 402–5 (405).

Bibliography

The following does not purport to be anything approaching a comprehensive bibliography, as comparison with the footnotes will show. It is, however, designed to be of maximum value to those who wish to explore in more detail some of the general themes developed in the book. For biographical references and special reading in matters of detail, see footnotes.

EDITED COLLECTIONS OF ESSAYS

Cardwell D. S. L. (ed.) *Artisan to Graduate, Essays to Commemorate the Foundation in 1824 of the Manchester Mechanics' Institution . . .* (Manchester University Press, 1974).

Cardwell D. S. L. (ed.), *John Dalton and the Progress of Science* (Manchester: Manchester University Press, 1968).

Crosland M. P. (ed.), *The Emergence of Science in Western Europe* (Macmillan, 1975).

Fox R. and Weiss G. (eds), *The Organization of Science and Technology in France 1808–1914* (Cambridge: Cambridge University Press, 1980).

Jordanova L. J. and Porter R. S. (eds), *Images of the Earth* (Chalfont St Giles: British Society for the History of Science, 1979).

Kent A. (ed.), *An Eighteenth Century Lectureship in Chemistry* (Glasgow: Jackson, 1950).

MacLeod R. and Collins P. (eds), *The Parliament of Science, The British Association for the Advancement of Science 1831–1981* (Science Reviews Ltd., 1981).

Mathias P. (ed.), *Science and Society, 1600–1900* (Cambridge: Cambridge University Press, 1972).

Musson A. E. (ed.), *Science, Technology and Economic Growth in the Eighteenth Century* (Methuen, 1972).

Rousseau G. S. and Porter R. (eds), *The Ferment of Knowledge: Studies in the Historiography of Eighteenth Century Science* (Cambridge: Cambridge University Press, 1980).

Teich M. and Young R. (eds), *Changing Perspectives in the History of Science* (Heinemann, 1973).

Turner G. L'E (ed.), *The Patronage of Science in the Nineteenth Century* (Leyden: Noordhoff International Publishing, 1976).

OTHER BOOKS

Allen D. E., *The Naturalist in Britain* (Pelican, 1978).

Ashton T. S., *The Industrial Revolution 1760–1830,* (Oxford: Oxford University Press, 1970).

Berman M., *Social Change and Scientific Organisation: The Royal Institution 1799–1844* (Heinemann, 1978).

Bernal J. D., *Science in History*, 4 vols. (Penguin, 1969).

Brockway L. H., *Science and Colonial Expansion: the role of the Royal Botanic Gardens* (Academic Press, 1978).

Cannon S. F., *Science in Culture: the early Victorian Period* (New York: Dawson and Science History Publications, 1978).

Cardwell D. S. L., *The Organisation of Science in England*, 2nd edn (Heinemann, 1972).

Clow A. and N., *The Chemical Revolution* (Batchworth Press, 1952).

Cohen I. B., *The Newtonian Revolution with illustrations of the transformation of scientific ideas* (Cambridge: Cambridge University Press, 1980).

Crosland M. P., *The Society of Arcueil* (Heinemann, 1967).

Crowther J. G., *Statesmen of Science* (Cresset Press, 1965).

Hahn R., *The Anatomy of a Scientific Institution: The Paris Academy of Sciences 1666–1803* (Berkeley, Los Angeles and London: University of California Press, 1971).

Hall A. R., *From Galileo to Newton 1630–1720* (Collins, 1963).

Hall A. R., *The Scientific Revolution 1500–1800* (Longmans, 1954).

Jacob M. C., *The Newtonians and the English Revolution 1689–1720* (Hassocks: Harvester Press, 1976).

Kargon R. H., *Science in Victorian Manchester* (Manchester: Manchester University Press, 1977).

Kelly T., *George Birkbeck, Pioneer of Adult Education* (Liverpool: Liverpool University Press, 1957).

Knight D. M., *The Nature of Science* (Andre Deutsch, 1976).

Lyons, Sir Henry, *The Royal Society 1660–1940. A History of its Administration under its Charters*, (Cambridge: Cambridge University Press, 1944).

Russell C. A. and Coley N. G. and Roberts G. K., *Chemists by Profession* (Milton Keynes: Open University Press, 1977).

Schofield R. E., *The Lunar Society of Birmingham* (Oxford: Clarendon Press, 1963).

Smith H., *The Society for the Diffusion of Useful Knowledge 1826–1846: a social and bibliographical evaluation* (Nova Scotia: Dalhousie Occasional Papers no. 8, 1974).

Stephens W. B., *Adult Education and Society in an Industrial Town: Warrington 1800–1900* (University of Exeter, 1980).

Thackray A. and Morrell J. B., *Gentlemen of Science: The Origins and Early Years of the British Association for the Advancement of Science* (Oxford: Oxford University Press, 1981).

Willey B., *The Eighteenth Century Background* (Pelican, 1972).

Wolf A., *A History of Science Technology and Philosophy in the Eighteenth Century,* 2nd edn (Allen and Unwin, 1958).

PAPERS IN ACADEMIC PERIODICALS

The following is a very small selection of papers which deal, in nearly every case, with broad and general issues only.

Ben-David J., 'The Profession of Science and Its Powers', *Minerva*, 10 (1972) 362–383.

Berman M., ' "Hegemony" and the Amateur tradition in British Science', *J. Soc. Hist.*, 1 (1975) 30–50.

Christie J. R. R. 'The Origins and Development of the Scottish Scientific Community, 1688–1760', *Hist. Sci.,* 12 (1974) 122–141.

Crosland M. P., 'The Development of A Professional Career in Science in France', *Minerva*, 13 (1975) 39–57.

Crosland M. P., 'Science and the Franco-Prussian War', *Soc. Stud. Sci.,* 6 (1976) 185–214.

Fox R., 'Scientific Enterprise and Patronage of Research in France 1800–1870', *Minerva*, 4 (1973) 442–73.

Fullmer J. Z., 'Technology, Chemistry and the Law in Early Nineteenth Century England', *Technology and Culture*, 21 (1980) 1–28.

Inkster I., 'The Social Context of an Educational Movement: A revisionist Approach to the English Mechanics' Institutes 1820–1850', *Oxford Rev. Educ.,* 2 (1976) 277–307.

Jensen J. V., 'The X-club, a social network of science in Victorian England', *Notes and Records Roy. Soc. London,* 24 (1970) 305–322.

Morrell J. B., 'Professors Robinson and Playfair, and the Theophobia Gallica: Natural Philosophy, Religion and Politics in Edinburgh 1789–1815', *Notes and Records Roy. Soc. London*, 26 (1971) 43–63.

Robinson E. and Musson A. E., 'Science and Industry in the late Eighteenth Century', *Econ. Hist. Rev.,* 13 (1960) 222–224.

Royle E., 'Mechanics' Institutes and the Working Classes, 1840–1860', *Hist. J.,* 14 (1971) 305, 321.

Shapin S. and Barnes B., 'Science, Nature and Control: interpreting Mechanics' Institutes', *Soc. Stud. Sci.,* 7, (1977) 31–74.

Stevens M. D. and Roderick G. W., 'Science, the Working Classes and Mechanics' Institutes', *Ann. Sci.*, 29, (1972) 349–60.

Thackray A., 'Natural Knowledge in Cultural Context: the Manchester Model', *Amer. Hist. Rev.*, 79, (1974) 672–709.

Wilde C. B., 'Hutchinsonian Natural Philosophy and Religious Controversy in Eighteenth Century Britain', *Hist. Sci.,* 18 (1980) 1–24.

Person Index

Subject Index